Mohammed Sghaïer Zaafouri

Reallocating natural ecosystems

Mohammed Sghaïer Zaafouri

Reallocating natural ecosystems

Tunisia's arid and desert zones

ScienciaScripts

Imprint
Any brand names and product names mentioned in this book are subject to trademark, brand or patent protection and are trademarks or registered trademarks of their respective holders. The use of brand names, product names, common names, trade names, product descriptions etc. even without a particular marking in this work is in no way to be construed to mean that such names may be regarded as unrestricted in respect of trademark and brand protection legislation and could thus be used by anyone.

Cover image: www.ingimage.com

This book is a translation from the original published under ISBN 978-620-6-70920-6.

Publisher:
Sciencia Scripts
is a trademark of
Dodo Books Indian Ocean Ltd. and OmniScriptum S.R.L publishing group

120 High Road, East Finchley, London, N2 9ED, United Kingdom
Str. Armeneasca 28/1, office 1, Chisinau MD-2012, Republic of Moldova, Europe
Printed at: see last page
ISBN: 978-620-7-98978-2

Contents

In memory of :

Supreme Combatant **Habib BOURGUIBA** first President of the Tunisian Republic. To this great man must be attributed the scientific level achieved by my generation and the generations of the 60s, 70s and 80s of the 20th century[ème] .

My dear mother, **Fatma**, who enrolled me in school in the early sixties of the 20th century[ème] despite the poverty and misery she lived in.

May God, the Merciful, bless them (4*^ψl 4ΛA>JJ^⅛A^-J)

SUMMARY

Reallocation is a technique applied to natural ecosystems that have reached the irreversible threshold (degraded or desertified ecosystems) at which restoration or rehabilitation fails to rebuild the original ecosystem. This technique is based on the introduction of allochthonous species and generally leads to a new ecosystem different from the pre-existing one, known as a "constructed ecosystem".

In Tunisia's arid and desert zones, under the arid Mediterranean and Saharan bioclimate, the degradation of natural ecosystems has reached the point of irreversibility, and restoration or rehabilitation techniques can no longer return these ecosystems to their initial state. As a result, recourse is often had to the technique of reafforestation using introduced allochthonous shrubs.

The bioclimatic limit for the successful reallocation of natural ecosystems in Tunisia's arid and desert zones is the lower understorey of the arid Mediterranean bioclimate. And the climatic one is an annual rainfall of at least 150 mm. These bioclimatic and climatic conditions must be met in very specific geomorphological situations and edaphic characteristics (plains, depressions, wadi beds, deep soils, low levels of active limestone and salts, *etc.*). However, these geomorphological situations and edaphic characteristics favourable to plant production are increasingly used for food crops (arboriculture and market gardening).

The success of the technique for reallocating natural ecosystems in arid and desert areas is generally problematic. The economic cost of installing and safeguarding introduced allochthonous shrubs is high. On the other hand, their success, survival and production remain very limited.

Keywords : arid and desert zones, Tunisia, reallocation, allochthonous shrubs, *Acacia saligna*

GENERAL INTRODUCTION AND ISSUES

Tunisia's natural ecosystems, particularly in **arid and desert areas**, have undergone profound changes since the mid-60s of the 20th century^{ème} , which have upset their dynamic balance. Attempts to reshape development policies since independence (1956), particularly in the agricultural sector, have had a remarkable impact on the population's way of life and consequently on the way natural resources (vegetation, water and soil) are used. The gradual and sustained shift in the population's lifestyle has had a clear impact on the way natural resources are exploited and used (Zaâfouri et *al.*, 2016).

The arid and desert zones covered by this study, referred to by Coque (1962) as **pre-Saharan Tunisia**, are located between the rainfall isohyets 100 mm/year and 200 mm/year.

The ecosystems of Tunisia's arid and desert zones were in **dynamic equilibrium** before the 1960s of the 20th century^{ème} . This dynamic equilibrium, which is both regressive and progressive, is due to the way in which natural resources are exploited and used. Until the middle of the 60s of the 20th century^{ème} , the only way of using the natural resources of the ecosystems in Tunisia, generally speaking, was pastoral farming and cereal growing, limited in time and space and practised occasionally. In these arid and desert areas, the nomadic lifestyle practised by a pastoral population and a small herd enabled the natural resources of the ecosystems to be reconstituted, more or less rapidly. Their dynamic equilibrium was then reflected in a relatively moderate degradation (regressive dynamics) due to grazing. After a few months' vegetative rest due to the transhumance of the herds, reconstitution (progressive dynamics), more or less rapid, was possible.

Several development policies have been applied in Tunisia by successive governments since independence (Fig. 1): from the **"absence of a clear policy"** until the beginning of 1960 to the **"liberal economy"** regime from the beginning of 1970, via the **"cooperative"** regime from the middle to the end of the 1960s (1964-1970), known by the Arabic name "**Taâdhed**". These attempts at development policies led to profound changes in the way people lived, and consequently in the way natural resources and ecosystems were exploited and used (Fig. 1).

The ecosystems of Tunisia's arid and desert zones during the 70s and 80s of the 20th century^{ème} remained in a state of threatened extinction, as defined by the IUCN (1997 and 1998), "**Vulnerable**". In other words, because of overgrazing and the destruction of their habitats by various uses, the ecosystems were at risk of falling into the categories of "**Endangered**" in the short term and "**Extinct**" in the medium term. And this is what happened from the 90s of the XXème century until the beginning of the XXIème century.

From the middle of the 1960s onwards, there was a general shift in the way people lived in Tunisia (Fig. 1): what had been a nomadic, pastoral population until the early 1970s became a peasant population, then a rural population (Attia, 1977), and finally a modern urban population from the beginning of the 21st century^{ème} (Zaâfouri et *al.*, 2016). This change in lifestyle gradually transformed the way in which natural resources were exploited: **from extensive** to **intensive farming** (Zaâfouri et *al.,* 2016). The consequences of this transformation were harmful to the dynamic balance and sustainability of natural ecosystems in Tunisia in general. Because of the fragility of ecosystems in arid and desert areas, the impact of these agricultural development policies was very noticeable on the natural resources of ecosystems (Fig. 1).

Dynamics of lifestyles, policies and operating systems

IJ	Period	1960 1S	>701980199020	OO 2020

	1964		1985	
Lifestyles	**Nomadic and pastoral population**		**Peasant and rural population**	**Urban population moving towards modernisation**
Agricultural development policies	**No clearly defined policy**	**Taâdhed" cooperative**	**"Liberal": State investment and subsidies to develop the agricultural sector and intensify agricultural production.**	**elopment**
Systems for exploiting natural ecosystems	**Pastoral** and episodic cereal farming	**Change to mixed farming: pasture and olive growing**	**Start of the creation of : Intensification of the creation of surface wells j of surface wells Development of ■ Intensification of irrigated irrigated market gardening market gardening : :: Creation of Drilling**	**Irrigated arboriculture and market gardening Reclaiming olive growing**
			1/2 century of intensive exploitation of natural ecosystems (1970-2020)	
State of natural ecosystems	**Dynamic balance**		**Start of disturbance from : Intensification of disturbance from dynamic equilibrium :: dynamic equilibrium : a/ endangered : a/ irreversibility threshold lowland ecosystems■ lowland ecosystems b/ Vulnerable : b/ Endangered other types of ecosystems: other types of ecosystems**	

Alteration of natural ecosystems

Figure 1: The relationship between the sustainability of natural ecosystems, changing lifestyles, farming systems and agricultural development policies in the Tunisian pre-Saharan region.

Based on Zaâfouri et *al* (2016), modified (2024)

Pastoralism, a very ancient heritage, became more and more sustained (overgrazing) and dry tree crops and rainfed cereal crops intensified (Fig. 1) following the sedentarisation and limitation of transhumance of the formerly nomadic population. The dynamic balance of natural ecosystems, which are fragile to say the least, is becoming increasingly fragile.

The consequences of the "**Taâdhed**" cooperative policy were disastrous for the sustainability of natural ecosystems, with a decline in acreage and a halt to progressive development.

This is the beginning of the qualitative and quantitative deterioration of natural ecosystems in Tunisia, particularly in arid and desert areas.

After the failure of the "**Taâdhed**" policy in 1970 of the late "super-Minister" Ahmed BEN SALAH (Minister: Public Health, Social Affairs, Planning and Finance and National Education, with a zealous right of oversight over the Ministry of Agriculture), a new regime was installed in the country. This was the era of the **"liberal economy"** of the late Prime Minister Hédi NOUIRA (19701980), which has left its mark on Tunisia to this day. During this decade, the main objective of this policy was to intensify food-producing agriculture in order to satisfy the food needs of a peasant and rural population threatened by famine and demographic growth.

The main aim of the development policy of this "powerful" Prime Minister and his successors was to improve the population's standard of living by intensifying agricultural production. This **"liberal"** policy was based on huge **investment** and **subsidies** by the State to develop the agricultural sector (livestock farming, arboriculture, cereal growing, irrigated crops). Unfortunately, the environmental aspect was neglected in this policy, and little importance was attached to it, as in all other economic sectors. The change in the use of natural resources, **deliberately** encouraged by successive governments unaware of the need for sustainable management of the environment and anxious to improve the standard of living of a poor and diminishing population, had contributed to the general deterioration of natural ecosystems in Tunisia.

The impact of agricultural development policies was very persistent on the natural ecosystems of Tunisia's arid and desert zones, which have long been vulnerable in terms of both natural vegetation and water and soil (Zaâfouri et *al.*, 2016). The degradation of these natural ecosystems is accentuated by climatic aridity (9 months/year to 12 months/year) and edaphic aridity (7 months/year to 9 months/year).

The policy of planning by the State and for the State (cooperativism), followed by the policy of intensifying agricultural production through State investment and subsidies, has gradually transformed Tunisia's natural ecosystems into "agro-ecosystem" farming systems.

Changes in both social organisation and farming and production systems, the consequences of policies applied without regard for the role of the environment in the context of sustainable development, have resulted in major disruptions to the dynamic balance of natural ecosystems in Tunisia's arid and desert zones (Chaieb, 1991; Auclair and Zaâfouri 1996; Picouet et *al.*, 1998, Tbib et *al.*, 2000; Chaieb and Zaâfouri, 2000. Zaâfouri et *al.*, 2016). The same is true of all the country's natural ecosystems.

From the middle of the 70s of the 20th century^ème (Fig. 1), state and, in particular, private irrigated perimeters multiplied in the spaces occupied by natural ecosystems: **this was the era of anthropisation proper**. The potentially fertile natural ecosystems of the plains fractured into meso-ecosystems and were gradually transformed into agro-ecosystems or "agar" systems (in the sense of Duvigneaud, 1974).

The dynamic balance of natural ecosystems that prevailed before the 1960s has been completely demolished since the end of the XXème century.

The intensification of subsistence farming, deliberately encouraged by the State, over half a century (1970-2020) has had harmful consequences for natural ecosystems (degradation and even disappearance). Lowland ecosystems have become **endangered** and some have become completely **extinct** (Fig. 1). Ecosystems subject to the forestry regime, those on communal land and on biologically non-productive marginal land, are gradually moving towards the status of **"Endangered"** (Fig. 1).

The anthropozoic pressure (extension of agriculture, clearing and overgrazing) exerted on Tunisia's natural ecosystems, in general terms, has become increasingly intense and sustained since the middle of the 70s of the XXème century: **this is the era of the extension of agro-ecosystems and the beginning of the regression of natural ecosystems.**

The intensification of agriculture, without regard for sustainable development, has led to the disappearance of large areas of natural ecosystems in Tunisia. Overgrazing by livestock on rangelands in arid and desert areas (natural ecosystems) has exacerbated the quantitative and qualitative degradation. These natural pastoral ecosystems, covering around 2.7 million hectares in the study area, can only produce $0.18.10^{8}$ UF/year (30 UF/ha/year), i.e. 15% of the needs of sheep, goats and camels, estimated at $5.28.10^{8}$ UF/year (Zaafouri et *al.*, 1994b). The fodder deficit is therefore 75%.

Accentuated by climatic and edaphic aridity, the impact of the intensification of subsistence farming and overgrazing has led, depending on the level of deterioration of ecosystems, to either degradation (reversible deterioration) or desertification (irreversible deterioration). Desertification is the consequence of agricultural practices that are not adapted to the climatic and edaphic conditions of a region.

Restoration, in the sense of Aronson et *al* (1995), of natural ecosystems in Tunisia was generally possible during the 60s and 70s of the 20th century^{ème} . However, this technique was not applied. The extensive deterioration of natural ecosystems (disappearance of plant cover and rarefaction, or even absence, of keystone species) since the end of this century means that this technique can no longer be applied. **Rehabilitation**, as defined by the same authors, remains the second possible technique for safeguarding natural ecosystems. Moreover, the arid climate and soil, the advanced stage of degradation, the slow rate of healing and the continued use by a population unaware of the environmental, economic and social role of natural ecosystems are the factors that limit the success of this technique in Tunisia, particularly in arid and desert areas (Floret et *al.*, 1983; Chaieb, 1989; Neffeti et *al.*, 1991; CEE, 1997). According to Aronson et *al* (1995), **the** only way of safeguarding natural ecosystems is to **use the reallocation technique**. However, this rehabilitation technique generally results in the creation of new ecosystems different from the original ones.

Below is a brief reminder of the concepts of **Restoration, Rehabilitation** and **Reallocation,** according to Aronson et *al* (1995), often referred to by restoration specialists as the "**3Rs**":

-Restoration: **"a** technique that aims to repair, as quickly as possible, the resilience and productivity functions of a damaged or simply blocked ecosystem, by repositioning it on a natural trajectory to return to its previous state".

This technique is generally used in ecosystems that **are in the process of being degraded**, simply by setting aside the land. It relies essentially on the natural dynamics (regeneration) of the ecosystem's flora. The presence of keystone species is decisive for its success.

Rehabilitation: **"a** technique based on intentional human intervention to restore, as far as possible, the full taxonomic composition, structure, diversity and dynamics of the pre-existing ecosystem considered to be indigenous and historical**".**

Applied to **degraded ecosystems**, this technique is based on strong human intervention: **(i)** the reintroduction of extinct native species and their associated microorganisms, **(ii)** soil work to improve hydric functioning, **(iii)** periodic maintenance and conservation work and **(iiii)** re-intervention in the event of failure or poor success.

-Reallocation: **"a** technique used when an ecosystem has been seriously degraded or desertified and it is impossible to return it to its original state, whatever the nature of human intervention".

Reallocation is based on the introduction of allochthonous species from other countries whose ecological zones are similar or almost similar to the areas to be reallocated. This technique generally results in a new ecosystem known as **a "constructed ecosystem"**. This new ecosystem has no structural or functional relationship with the original ecosystem.

It should be noted that in certain cases and depending on the country's general development policy, natural ecosystems, even if not degraded, could be reallocated for other uses: industrial, urban, tourism, *etc.*

Most of the ecosystems in Tunisia's arid and desert zones have reached, or even exceeded, the threshold of irreversibility. The extent of degradation, to the point of desertification, the high cost and the failure of Restoration and Rehabilitation techniques have led Tunisia's agricultural development policy in the environmental field to focus on **the Reallocation technique.** This technique began with the introduction of shrubs. ème*Opuntia ficus-indica* was introduced into Tunisia in the early decades of the 20th century (1920-1930) (Manjauze et le Houérou, 1965). Shrubs of the *Atriplex* genus were introduced in the 1940s (Franclet and Le Houérou, 1971). By 1965, shrubs of the *Acacia*, *Prosopis*, *Parkinsonia* and *Eucalyptus* genera had spread throughout the country (Le Houérou and Pontanier, 1987; Akrimi and Zaâfouri, 1989; Zaâfouri, 1989; Zaâfouri et *al.*, 1991; Zaâfouri, 1993; Zaâfouri et *al.*, 1994b, 1995 and 1997).

In arid and desert areas of Tunisia, reafforestation was mainly carried out using shrubs of the *Acacia*, *Atriplex*, *Prosopis*, *Parkinsonia* and *Eucalyptus* genera (same authors).

The reallocation of Tunisia's natural ecosystems has led to new plant formations based on non-native species, thus creating new ecosystems that are very different from the original ones. These "constructed" ecosystems are referred to by pastoralists as sylvo-pastoral perimeters, by foresters as sylvopastoral plantations and by ecologists as sylvo-pastoral ecosystems.

More often than not, allochthonous shrubs were introduced into Tunisia's natural ecosystems without any prior knowledge of their ecological requirements, which were, moreover, poorly known even in their countries of origin (Boudy, 1950). On the subject of **acacias** introduced into Tunisia, this author wrote: "there was a complete lack of precise information on the temperament and ecology of these species in their countries of origin". He added: "All these acacias are, moreover, scorned and consequently poorly studied, even in their countries of origin".

This remark applies to all the non-native shrubs introduced into Tunisia since the beginning of the 20s of the XXème century. Although this remark dates back approximately 3/4 of a century, we unfortunately continue, until today (2024), to use the same shrubs in the reallocation of Tunisia's natural ecosystems, whatever their ecological aptitudes and biological potential.

Native shrubs such as *Periploca laevigata*, *Rhus tripartitum*, *Ceratonia siliqua*, *Pistacia atlantica* and *lentiscus*, *Acacia tortilis*, *Atriplex halimus*, *Ephedra alata* and *altissima*, *Calligonum arich* and *azel*, *etc.*, which are better adapted to the ecological conditions of arid and desert zones, were and are almost completely neglected.

The low level of knowledge of the ecological requirements of shrubs introduced into Tunisia, which are poorly understood even in their natural areas, was certainly one of the causes of many failures in the reallocation of Tunisia's ecosystems, especially in arid and desert areas. The simple observation of a few interesting biological characteristics (rapid growth and high biomass), disregarding all ecological and environmental considerations, had favoured their use and neglected indigenous shrubs with slow growth and low biomass.

The environmental impact of some of these introduced shrubs, such as the allopathic effect on the native flora, was never taken into consideration. The disappearance of the natural flora under the majority of these shrubs and, in particular, under the genus *Eucalyptus* is a perfect example of this allopathic impact. In addition, other shrubs are becoming invasive, competing with the native flora (*Prosopis* ssp. for example).

the main allochthonous shrubs introduced for the reallocation of natural ecosystems in Tunisian arid and desert zones are *Acacia saligna*, known in North Africa under the erroneous nomenclatural combination of *Acacia cyanophylla*, *Acacia cyclops*, *Acacia ligulata*, *Parkinsonia aculeata*, *Atriplex nummularia*, *Prosopis* ssp, *Eucalyptus* ssp. (Hamza, 1986; Akrimi et Zaâfouri, 1990; Zaâfouri, 1989; Zaâfouri et *al.*, 1991; Zaâfouri, 1993 and Zaâfouri et *al.*, 1991, 1994b, 1995 and 1997).

The aim of these introductions was to improve the biological production of the ecosystems and create standing pastoral reserves. The reactivation of the progressive dynamics of natural ecosystems was the manager's wish.

Although some spectacular successes have been described for certain shrubs introduced under particularly favourable environmental conditions (Schönenberger, 1971; El Hamrouni and Sarson, 1975a; Karem, 1984; Karem and Ktata, 1986; Le Houérou and Pontanier, 1987; Zaâfouri, 1989 and 1993, Neffeti et *al*, 1990; Zaâfouri, 1993; Zaâfouri et *al.* 1991, 1994b, 1995 and 1997); however, it must be acknowledged that there have been many spectacular failures (same authors).

In this book, we analyse the response of one of the allochthonous shrubs most commonly used in the reallocation of ecosystems in Tunisian arid and desert zones: ***Acacia saligna***. We attempt to provide some answers to the threshold limits of the tolerance of this shrub to the ecological conditions of these zones and the limits of the reallocation technique.

Finally, this work was the subject of a Diplôme d'Etudes Approfondies dissertation in Mediterranean Ecology and the first part of a doctoral thesis defended respectively in 1989 and 1993 at the Université de Droit, d'Economie et des Sciences d'Aix- Marseille, France (Faculté des Sciences et Techniques de Saint Jérôme de Marseille). These dissertations are complemented by our research work since 1994 in the field of reallocation of natural ecosystems under the arid Mediterranean and Saharan bioclimate.

This book is divided into three chapters:

- the first describes *Acacia saligna* and discusses the evaluation approach;
- the second characterises the ecological factors of Tunisia's arid and desert zones and reaffected ecosystems;
- and the third chapter looks at the response of this allochthonous shrub to the ecological conditions of the reallocated ecosystems and, consequently, the success or failure of the reallocation technique using introduced shrubs.

Chapter 1

ALLOCHTHONOUS SHRUBS USED FOR REALLOCATION AND THE EVALUATION APPROACH

1. NON-NATIVE SHRUBS USED FOR REALLOCATION

1.1. THE IMPORTANCE OF SHRUBS IN ECOSYSTEM RESTORATION

In the arid and desert areas of Tunisia, where the bioclimates are Mediterranean: lower arid and upper Saharan (average rainfall of 100 mm/year to 200 mm/year), the allochthonous shrubs used to rehabilitate natural ecosystems were introduced in the mid-1960s at the height of the "Taâdhed" cooperative policy. The first shrubs used in this area were *Acacia ligulata* and *Opuntia ficus indica.*

Acacia saligna (=Acacia *cyanophylla*) was not regularly used until the early 1970s, replacing *Acacia ligulata*, which proved unpalatable to animals, and *Opuntia ficus-indica*, which is not very resistant to drought with rainfall of less than 150 mm/year. From the 1980s onwards, the reallocation of natural ecosystems in arid and desert areas of Tunisia by introduced shrubs, mainly *Acacia saligna*, intensified. Other allochthonous shrubs of the *Acacia*, *Prosopis*, *Atriplex* and *Parkinsonia* genera appeared sporadically.

The main objective of introducing these shrubs was, as we have just discussed in the general and problematic introduction, to improve the biological production of natural ecosystems, the creation of standing pastoral reserves and to hope for the reactivation of the natural dynamics of indigenous vegetation.

The reallocation of Tunisian arid and desert ecosystems was based on the use of around ten introduced shrubs (Fig. 2). Almost all the reallocated ecosystems (97.5% of the areas) were based on : *Acacia saligna*, *Acacia ligulata*, *Acacia salicina*, *Acacia cyclops*, *Prosopis juliflora*, *Prosopis chilensis*, *Parkinsonia aculeata*, *Atriplex nummularia*, *Atriplex canescens*, *Opuntia ficus-indica and Eucalyptus* ssp. Native shrubs (*Atriplex halimus*, *Calligonum azel*, *Periploca laevigata* and *Ceratonia siliqua*) account for only 2.5% (Fig. 2).

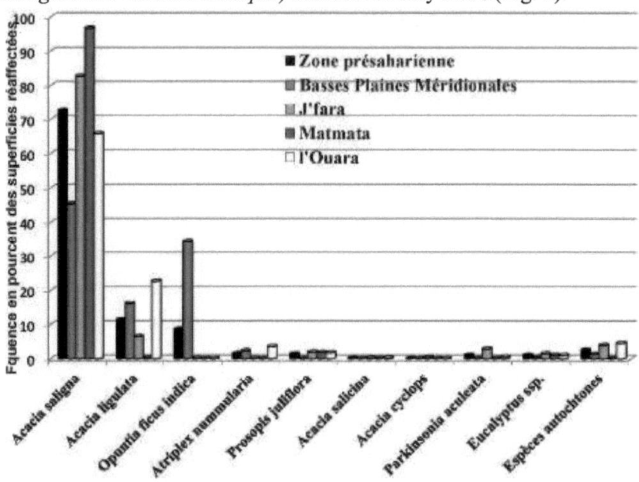

Figure 2: Importance of allochthonous shrubs (in % of surface area) in the reallocation of natural ecosystems in arid and desert areas of Tunisia.

According to Zaafouri (1993, completed 2020)

The wide range of allochthonous shrubs used in the reallocation of natural ecosystems in Tunisia, in general, can be explained by the absence of perfect knowledge of their ecological requirements. **It's more a case of trial and error than a technique based on the results of scientific experimentation.** [ème]Unfortunately, Tunisia's natural ecosystems continue to be reallocated by the same shrubs, despite the results of scientific studies carried out since the last quarter of the 20th century (1975), which have shown the limitations of some of these shrubs introduced into arid and desert areas (see Bibliography).

Among the allochthonous shrubs introduced into Tunisia, *Acacia saligna*, family *Fabaceae* (*Leguminosae*), is the cornerstone of the reallocation of Tunisia's natural ecosystems. In arid and desert areas, this shrub occupies an average of 72.5% of reallocated areas (Fig. 2). In several ecosystems, it occupies an area of 65% to 97% (Fig. 2). The primary objective in choosing *Acacia saligna* for reallocation is fodder production (high phytomass) to make up for the fodder deficit in rangelands. Its nodulation capacity (Mansouri, 2011) and the symbiotic process of its microorganisms (Amrani et *al.*, 2009) could enrich the soil with nitrogen and reactivate the natural dynamics of the native flora.

1.2. Characteristics of *Acacia saligna*

Acacia saligna (Labili.) H.L. Wendl. is generally known in North Africa under the erroneous nomenclature combination *Acacia cyanophylla* Lindl. Bayley (1914) in Léone 1924 considered *Acacia cyanophylla* Lindi. synonymous with *Acacia saligna* (Labill.) H.L. Wendl. Passavalli (1935) distinguished *Acacia saligna* from its synonym *Acacia cyanophylla* by the colour of the leaves, green-azure (verde-azzuro), and the more visible lateral veins (nerviture laterali più visibili).

1.2.1. Systematic classification

Two types of systematic classification are commonly used: **(i)** the classical or Linnaean classification and **(ii)** the phyllogenetic classification. We have chosen the Angiosperm Phylogenetic Group classification (Angiosperm Phylogeny Group, 2003 and 2009).

According to this type of classification, the simplified systematics of *Acacia saligna* is as follows (Tab. 1)

Table 1. Simplified taxonomy of *Acacia saligna* according to the phylogenetic classification (APG, 2003 and 2009)

Domain or Empire	*Eucaryota*
Kingdom	*Plantae*
Subdomain	*Tracheobionta*
Division	*Magnoliophyta*
Class	*Angiosperma*
Subclass	*Tricolpata*
series	*Rosidae*
Sub-series	*Faboideae*
Order	*Fabales*
Family	*Fabaceae*
Subfamily	Mimosaceae (=Mimosoidae)
Type	*Acacia*
Species	*saligna* (=*cyanophylla*)

1.2.2. Botanical and morphological characteristics. Food and energy values

We based our botanical and morphological description of *Acacia saligna* and its food and energy value on the work carried out by Vassal (1972), Le Houérou and Ionesco (1973), El Hamrouni and Sarson (1974), Monod (1974), N.A.S, (1980), Sarson and Salmon (1977), Le Houérou (1980), Dumancic and Le Houérou (1981), Akrimi (1986), Chouaieb (1982), Dahl (1982), Le Houérou and Pontanier (1987), Akrimi and Zaâfouri (1990), Zaâfouri (1993) and many others.

Acacia saligna (Fig. 1A) is generally a very polymorphous shrub 2 to 6 m high, forming an extremely heterogeneous stand. In its natural range, it can be as tall as a 9 m tree. It has a sinuous, highly branched form with a short trunk (20 to 80 cm). The bark on the trunk is light green, smooth when young and cracked and brownish-grey when mature. This year's twigs are smooth and light green, while older twigs are greenish and more or less fissured.

Photo 1. ***Acacia saligna*** (Labill.) H.L. **(=** ***Acacia cyanophylla*** Lindl.**)** **A:** habit **B:** phyllodes **C:** flowers **D:** fruit (pods)

The first pairs of leaves on *Acacia saligna* and all species in the *Acacia* genus are pinnate or bipinnate compound leaves. These true leaves disappear a few days after germination. They are replaced by **phyllodes** (Photo 1B), which are slightly falcate with a well-marked main vein, and are greenish-blue in colour; hence the erroneous nomenclature *Acacia cyanophylla* and the common French name "Mimosa ou Acacia bleuâtre".

The flowers (Photo 1C), 1 cm to 1.5 cm in diameter, are arranged in glomerules forming clusters (sometimes solitary). The flowers are bright yellow to orange-yellow and are located at the tips of the branches.

The fruits, in the form of pods (Photo 1D), are long, flat, fairly narrow between the seeds, about 3 cm to 7 cm long and 0.5 cm wide. They are slightly arched and not very tortuous, greenish in colour during ripening and blackish at maturity. Each pod contains several greenish seeds before ripening and they become completely black when ripe.

Acacia saligna is an extremely heterogeneous species in terms of habit (from drooping to rectilinear) and the shape and size of the phyllodes. Zaâfouri (1993) highlighted the great variability in the size of the phyllodes in this species in stands in arid and desert areas of Tunisia. He distinguished several types of phyllodes on the basis of the length/width ratio: **"phyllodes latifolia"** (length/width ratio < 3) to **"phyllodes longifolia"** (length/width ratio > 40). Between these two types of phyllodes, there are around ten dimensions (Zaâfouri, 1993).

Acacia saligna is a species that multiplies easily by seed and has the ability to regenerate perfectly at any cutting height (stump or trunk regeneration) and to sucker from the roots (root suckering). Although this species is considered invasive, the dry period in arid and desert areas beginning just after the wet season is the limiting factor for the survival of young plants resulting from regeneration by seed (Zaâfouri, 1993).

The phyllodes, young shoots and fruits of *Acacia saligna* are palatable to animals. They are eaten by sheep, goats and even cattle and camels (personal observations). However, palatability differs from one plant to another and from one type of phyllodes to another: the "longifolia" type is more palatable than the "latifolia" type (Zaâfouri, 1993).

Phyllodes are rich in protein: 16% crude protein and 12% digestible protein per kilogram of dry matter. The digestible nitrogen content and energy value (Feed Unit FU) are, successively, 140 g/kg DM and 0.31 FU/kg DM. The in vivo digestibility of the organic matter is around 55%, a relatively high value for a woody forage.

1.2.3. Origin, geographical distribution and environmental conditions of the natural range

Native to Western Australia (Maslin and Pedley, 1982; Maslin, 2001a b and 2002), *Acacia saligna* was introduced to South Africa in 1840 for dune fixation (N.A.S., 1980). It has also been introduced in Uruguay, Mexico, Palestine, Jordan, Iraq, Iran, Greece, Chibre (N.A.S., 1980) and France (Vassal, 1984).

In North Africa, *Acacia saligna* appears to have been introduced for the first time in Algeria in the early 1870s (Bert, 1886 in Léone, 1924). It appears to have been introduced on a large scale in North Africa in Libya in 1916 to fix the coastal dunes of Tripoli (Léone, 1924).

In Tunisia, according to Le Houérou and Pontanier (19987), *Acacia saligna* was introduced in the 1930s in the Bizerte region to fix the maritime dunes. ème It began to spread beyond the coastal regions in 1965 and was widely distributed from the 1980s onwards (Akrimi and Zaafouri, 1990; Zaâfouri, 1993 and Zaâfouri et *al.*, 1994b).

The environmental conditions in the natural range of *Acacia saligna*, Western Australia, described by the authors cited above are :

- Average annual rainfall: 600 mm to 1000 mm ;
- summer temperature: 23°C to 36°C ;
- winter temperature: 4°C to 9°C ;
- temperature range: 19°C to 27°C ;
- optimum summer temperature: 30°C ;
- optimum winter temperature: 13°C ;
- altitude: from ocean level to 300 m above sea level. At high altitudes, it occurs in isolated, stunted individuals;
- Soil: a psamophilous species typical of more or less fixed coastal dunes. It fears acidic soils, limestone-rich soils and salty soils.

Although the thermal and edaphic conditions in the natural area of *Acacia saligna* are almost similar to those in the arid and desert zones of Tunisia (see Chapter 2), the rainfall conditions (600 mm/year to 1000 mm/year) are too far away (100 mm/year to 200 mm/year). However, the rainfall conditions (600 mm/year to 1000 mm/year) are too far away (100 mm/year to 200 mm/year). In Algeria, a Mediterranean bioclimate, Khéloufi (2019) reported that the lower rainfall limit for this shrub is 250 mm/year.

2. THE EVALUATION APPROACH

We opted for **a survey approach**, also known as a **comparative approach or method**, to assess reallocation through the introduction of *Acacia saligna* in Tunisian arid and desert zones.

The principle of this approach is to select a set of ecosystems covering a range of climates and soils (Garbaye et *al*., 1970 and le Tacon, 1973). This method is based on **the level of perception of the ecosphere**, as defined by Tansley (1935), and on **the notion of ecosystem** as defined by Evans (1956).

According to Garbaye et *al* (1970), the survey approach or comparative method is generally very tricky to implement and difficult to tackle because of the constraints imposed by the environment and the stand. Among these constraints, the authors mention

- climatic heterogeneity ;
- geomorphological heterogeneity ;
- edaphic heterogeneity ;
- stand heterogeneity

However, ecological diagnosis based on the perception of the ecosphere, as defined by Godron and Poissonnet (1972), Duvignaud (1974) and Long (1974 and 1975), could solve the problems posed by the heterogeneity of the environment and the population.

The level of perception, **ecosystem (ecological system)**, a concept widely used in ecological studies, is the basic unit of ecological diagnosis (authors quoted above). According to these authors, the ecosystem is at the base of the level of perception pyramid (see diagram below). It should be noted, however, that the level of perception varies from one author to another depending on the objective they have set.

1st level → Ecosphère

↓

2ème level → Ecological regions

↓

3ème level → Ecological sectors

↓

4ème level → Ecological systems (ecosystems)

2.1. THE SURVEY APPROACH

The survey-based approach to spatial diagnosis involves three successive stages:

- space zoning ;
- the choice of sampling system ;
- the choice of environmental and species variables.

2.1.1. Space zoning

Our study relates to the action of ecological factors on the establishment, success and production of *Acacia saligna*, the key importance being, therefore, the homogeneity of climatic, geomorphological, edaphic and floristic factors.

To achieve this, we have adopted the following approach to zoning the Tunisian arid and desert areas:

- choice of ecosphere ;
- selection of ecological regions ;
- delimitation of ecological sectors ;
- delimitation of ecosystems.

2.1.1.1. Choice of ecosphere

The ecosphere is a concept created by the American ecologist Lamont Cole in 1958 to designate a set of ecosystems in which living organisms interact sustainably with each other and with the environment. The elements that make up the ecosphere are inseparable.

Taking this concept into account, we have considered the arid and desert zones of Tunisia, **pre-Saharan Tunisia** according to Coque (1962), as **an ecosphere**. These areas lie between the 100 mm/year and 200 mm/year rainfall isohyets (Fig. 3).

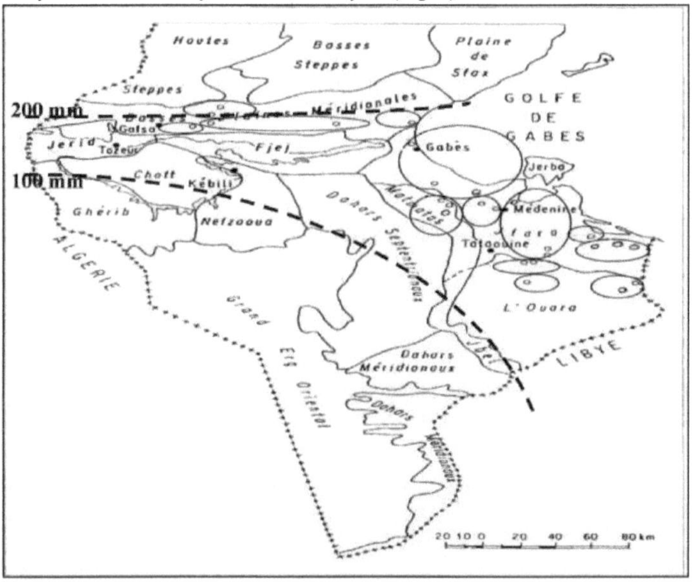

Figure 3: Zoning of Tunisian arid and desert areas into ecological regions, ecological sectors (large circles) and ecosystems (small circles).

Approximate rainfall limits for the arid and desert zones studied (100 mm/year-200 mm/year)

Geographically, it lies between parallels 31°59' and 34°30' North and meridians 7°20' and 11°50'. Or its geographical limits, according to this author, are :

-From North to West: the imaginary line linking the Mediterranean Sea, at the town of Mahrès, to the Tunisian-Algerian border, passing through the Gafsa mountain range (jbels: Mazzouna, Bou Hedma, Orbata, Ben Younes, Bou Ramli) and the towns of Om Al Arayes and R'deyef.

-from West to South: the boundary is also marked by an imaginary line linking the Tunisian-Algerian border to the Tunisian-Libyan border, passing through Tamaghza-Chott Jérid-Kébili-Gsar Ghilène-Borj Bourguiba as far as Dhhiba.
The arid and desert areas of Tunisia were chosen as part of a national research programme to restore, rehabilitate and reallocate degraded and desertified ecosystems. The Institut des Régions Arides Médenine, Tunisia, a state research body, is the project manager for this programme. As a teacher-researcher in this institution, I was in charge of **the ecosystem reallocation** sub-programme.
According to Coque (1962), Le Houérou (1959 and 1969), Floret and Pontanier (1982 and 1984), Saghiri (1991), these areas are characterised by: **(i)** a climate that varies according to the gradient of latitude and longitude and a degree of continentality, **(ii)** a very varied range of geomorphological situations (jebels, glacis, plains, depressions, wadi beds), **(iii)** different soil types and **(iii)** diversified plant groupings.
These arid and desert zones, retained as an ecosphere, are heterogeneous in terms of climate, geomorphology, soil and vegetation. However, they constitute an ecological entity characterised by the same climatic (9 months/year to 12 months/year) and edaphic (7 months/year to 9 months/year) aridity and by the same type of vegetation (steppe vegetation).

2.1.1.2. Selection of ecological regions

In order to minimise the heterogeneity of ecological factors in this ecosphere as much as possible, the arid and desert zones of Tunisia are fragmented into **ecological regions**. Within each region, climatic conditions (rainfall and temperature) are more or less homogeneous. However, geomorphological, edaphic and vegetation type heterogeneity is still significant.
For our purposes, **the ecological regions** are **the natural regions of southern Tunisia** delimited by Le Houérou in 1959 (Fig. 3). Nine (9) natural regions were defined by this author. However, we have only considered four (4) regions (Fig. 3 and Tab. 2):

- **the Southern Lowlands ;**
- **the Matmata ;**
- **la J'fara ;**
- **the Ouara**.

Table 2. Spatial zoning results for arid and desert zones in Tunisia

Natural region (Le Houérou, 1959) Ecological region by design	Number of	
	Ecological sectors	Ecological systems
Southern Lowlands	4	6
Matmata	1	3
J'fara	5	12
Ouara	3	5
Jérid	small areas reallocated analysis reliable statistics	
Féjij		
Steppe Dhahar		
Southern Dhahar		
Nefzaoua		

The fact that only 4 ecological regions were selected from a large area was dictated by the

surface areas reallocated by the introduction of allochthonous shrubs. In fact, after surveying all the arid and desert zones, we noticed that in the rest of the natural regions (ecological regions), the areas reallocated were very small (a few tens of square metres), making it impossible to draw any conclusions from the results.

The main climatic conditions (rainfall and temperature) are as homogeneous as possible in each of these ecological regions. However, geomorphological, edaphic and floristic heterogeneity remains significant.

2.1.1.3. Delimitation of ecological sectors

The aim of delimiting ecological sectors is to homogenise geomorphological, edaphic and floristic factors. Each ecological region is therefore subdivided into **ecological sectors**. Two parameters are used to define the ecological sector:

-the location on the general toposequence (geomorphology),

-the type of natural vegetation (ecological group).

The area reallocated within an ecological region is, according to this approach, either **a plain**, **a glacis**, **a depression** or **a wadi bed**. The number of ecological sectors depends on the surface area reallocated within the ecological region (Fig. 3 and Tab. 2).

2.1.1.4. Defining ecosystems

At the level of perception, the ecological sector, the climatic and geomorphological factors are almost homogeneous. There remains, however, a heterogeneity of edaphic and floristic characteristics. In order to reduce, but not eliminate, edaphic and floristic heterogeneity, we have subdivided the areas reallocated in each ecological sector into **ecosystems**. At this level of perception, all the ecological factors are as homogenous as possible (climate, geomorphology, soil and flora).

The number of ecosystems delineated in each ecological sector is a function of the area reallocated (Fig. 3 and Tab. 2).

The ecosystem can be identified and characterised by two concepts: **the ecological station** or **the forest station**:

- **The ecological station:** Rol (1954) defines the ecological station as "an area of land, which varies greatly in surface area, from a few dcm^2 to several km^2 , but which is homogeneous in terms of the ecological conditions which prevail there. In other words, the ecological station is a topographical unit defined by a set of climatic, edaphic and biotic factors". It is also defined by Bonneau and Timbal (1973) as "an intuitive biological unit that corresponds to the perception of a plant landscape that is homogeneous in its composition and occupies a well-defined topographical position".
- **Forest station:** defined by Decourt (1973 a) as "an area of forest with homogeneous ecological conditions and stands in which the forester can practise the same silviculture and expect the same production".

In the concept of ecological station, the notions of biomass, energy balance and dynamic and trophic relationships between the various elements of the biocoenosis, although essential in the concept of ecosystem, are not explicitly included in the above definitions of ecological station. It should be noted that the notion of ecosystem is dynamic and energetic and generally refers to its functioning and production (Evans, 1956 and Duvignaud 1974). On the other hand, in the notion of forest station, we note that it explicitly closes off the notions of biomass and energy balance (production), which are essential for assessing reallocation.

In our conception of the ecosystem, and for the purposes of this study, it is closer to the notion of a forest station than to that of an ecological station.

Whether it is an ecological station or a forest station, the level of perception, ecosystem, in addition to the homogeneity of climatic factors and geomorphological conditions, is characterised by the homogeneity of edaphic characteristics. The phytocoenosis is also homogeneous (same ecological group). This level of perception of the ecosphere (ecosystem) is therefore characterised by the homogeneity of all ecological factors.

The delimitation of the ecosystem must be rigorous in order to guarantee the homogeneity of ecological factors, mainly edaphic factors. To achieve this, we have selected a number of parameters:

- the nature of the bedrock or substrate;
- the nature of the soil ;
- the nature and thickness of the loose layer ;
- the presence or absence of wind cover ;
- water infiltration capacity ;
- the plant group or ecological group.

The number of ecosystems delimited is a function of the area reallocated to each ecological sector (Fig. 3 and Tab. 2).

Thus, the zoning of the space of the reallocated arid and desert areas discriminated 4 regions, 13 sectors and 26 ecological systems (Fig. 3 and Tab. 2). The number of sectors in each region and ecosystems in each sector is a function of the size of the reallocated area.

This approach has made it possible to standardise :

- **climatic factors in each ecological region ;**
- **climatic and geomorphological factors in each ecological sector**

-all the ecological factors (climate, geomorphology, soil and phytocoenosis) in each ecosystem.

2.2. Sampling and plot selection

2.2.1. Type of sampling

Our objective is to explain the causes of success or failure of the introduced shrub, *Acacia saligna*, in the reallocation of ecosystems in Tunisian arid and desert zones. Consequently, sampling must be adequate to achieve this objective.

Several sampling methods have been proposed (Pardé, 1961; Lamotte, 1962; Gounot, 1969; Garbaye et *al.*, 1970; Peters, 1974; Floret, 1988; Zaâfouri et *al.*, 1974a). According to these authors, each type of sampling has its advantages, disadvantages and limitations.

For our study, we **practised two-stage sampling using multistage sampling** (authors above). This type of sampling is neither systematic nor random: **it is exploratory.**

This approach is based on a survey in search of areas reallocated by allochthonous shrubs, with the aim of diversifying to cover as much as possible of each ecological region in terms of geomorphology and lithology. The result is either **over-** or under-representation of the sectors in each region and of the ecosystems in each sector.

2.2.2. Choice of plots: area and number

The demonstration of the influence of ecological factors on the success or failure of *Acacia saligna*, used for the reallocation of these ecosystems, requires plots with ecological conditions, mainly edaphic conditions, that are as homogeneous as possible. In this case, however, it is difficult to use plots with a constant surface area as defined and proposed by Pardé (1961). If the plot area is small, the number of individuals may be too small for a reliable statistical analysis. Otherwise, there is a risk of edaphic heterogeneity. **The objective**

can therefore only be achieved by plots of varying size.

We have thus considered as the elementary surface of the plot that occupied by the 15 most neighbouring individuals in the ecosystem delimited in advance (Cf. § 2.1.1.4 above). Pardé (1961) recommended, in forest inventory, 10 to 20 individuals per plot in order to be able to carry out a reliable statistical analysis.

The number of plots selected depends on the size of the area reallocated in the ecosystem. This number varies from 2 plots to 3 plots. Thus the total number of individuals in an ecosystem is 30 individuals or 45 individuals. This number therefore allows a credible statistical analysis.

2.2.3. Choice of environmental and species variables

In this work, we aim to explain the behaviour of the *Acacia saligna* shrub in the different ecological situations of the ecosystems of the Tunisian arid and desert zones and, consequently, to evaluate its reallocation (success or failure).

To meet this objective, we carried out two types of observation:

-Observations relating to the ecosystem: environmental variables (explanatory variables) ;

-observations relating to the species (explained variables).

2.1.3.1. Environmental variables: Explanatory variables :

These are variables relating to the edaphic characteristics of the ecosystem. In general, the soil type integrates all the edaphic parameters. There are many soil variables that can be measured or observed, but their value varies (Bonneau and Timbal, 1973). We have therefore selected those soil parameters that have a remarkable influence on the success and production of *Acacia saligna*. These parameters are mainly :

- genetic characteristics of the soil and parent rock ;
- physical characteristics of the soil: nature and thickness of the loose layer, water infiltration and storage capacity (water regime), presence or absence of a water table and presence or absence of a wind cover;
- chemical characteristics of the soil: pH, organic matter, limestone and gypsum content and degree of salinity ;
- soil types according to the French classification.

2.1.3.2. Species variables: Explained variables

To assess the response of *Acacia saligna* to the edaphic characteristics of the ecosystem, we combined two approaches: **estimation and evaluation:**

A/ <u>Valuation</u>

This approach consists of evaluating the behaviour of *Acacia saligna* using a qualitative assessment scale proposed by Schönenberger (1971). This scale has 5 levels: **1**: poor condition, **2**: poor condition, **3**: average condition, **4**: good condition and, finally, **5**: very good condition. Because of the subjectivity of this scale, we ourselves (Zaâfouri et *al.*, 1994b) improved it by introducing quantitative data and clarifying the meaning of qualitative data, as follows:

(1) poor condition: the success rate is very low (< 20%). Individuals are very sickly. Foliage is very sparse, yellow in colour or completely absent. Theoretically, the species cannot survive in the short term.

(2) poor condition: the success rate is low (20% to 40%). Individuals are puny. Foliage is generally green but sparse. Growth and production of the species are almost nil, but it manages to survive in the short term.

(3) average condition: the success rate could be either low (20 to 40%) or average (40 to 60%). The plants are moderately vigorous. Foliage is green and fairly abundant. The growth of the species, which is not negligible, allows it to survive in the medium term, but production is low.

(4) good condition: the success rate is sometimes low to average (20% to 50%). The plants are vigorous. Foliage is abundant and in good condition. The growth of the species is good and we can expect some production (average to good).

(5) very good condition: the success rate may sometimes be low (< 40%), but is generally high (> 60%). The plants are very vigorous, with foliage in excellent condition and very abundant. Growth and production are at their optimum.

It should be noted that the low success rate in scales 3, 4 and 5 for certain ecosystems is due to failures in the installation phase.

B/ Assessment

We considered two parameters to assess the success or failure of *Acacia saligna*: **(i)** the success rate or survival rate and **(ii)** production.

-The success rate or survival rate: this is the percentage of individuals surviving on the day of assessment in relation to the total number of individuals planted (planting density). Depending on the ecological region, planting density varies between 800 plants/ha and 1000 plants/ha (regional forestry districts).

-Production: this is a synthetic parameter of a plant's response to the ecological conditions of its environment. It incorporates all the plant's development and growth parameters.

The production considered in this study is **the phytomass of the aerial part**. This is defined by Feuillas (1979) as: "the quantity of plant mass produced by the aerial part of the species in a given time". It is either **a total phytomass** or **a phytomass per foot** (Clément et Touffet, 1976 and Feuillas, 1979). Total phytomass is the quantity of plant matter on the individual (above-ground phytomass) and the dead phytomass rejected on the ground (leaves, branches, twigs, fruit, *etc.*).

The phytomass of the aerial part, also known as the **epigeous biomass** (Feuillas, 1979), is generally expressed as dry matter (DM) per unit area and per unit time. It can also be expressed as green matter (GM).

Because of the difficulty of assessing total phytomass due to the lack of data on phytomass released to the soil since the date of planting (4 to 6 years), we have limited ourselves to standing phytomass or biomass removed from the soil.

Several methods for assessing the phytomass of shrub species have been described (Feuillas, 1979; Etienne, 1980; Barbéro, 1981; Asocar et *al.*, 1981; Felker et *al.*, 1982; FAO, 1982; Baudin, 1985; Floret et *al.*, 1983; Floret, 1988; Zaâfouri, 1993 and Zaâfouri et *al.*, 1994a). These methods are either **(i) destructive, (ii) semi-destructive** or **(iii) non-destructive**. Each method has its advantages and disadvantages (see authors cited above).

Because of the conditions imposed by the forestry services in the study area (ban on logging shrubs), we tried to find a compromise between the semi-destructive and non-destructive methods.

We therefore chose the method of assessing the phytomass removed by measuring the biovolume, with the felling of a single individual per plot. This individual is the average biovolume individual, by analogy with the average volume stem used in forestry (Prodan, 1951). The individual of average biovolume is defined, as in forestry: **the individual whose**

biovolume multiplied by the number of individuals in the plot gives the total biovolume. Choosing the tree of average biovolume comes down, on the one hand, to the conditions imposed by the forestry services and, on the other hand, to minimising the effect of plantation density and the success rate on production. In fact, the production of the average biovolume tree is very clearly little influenced by these two parameters (Zaâfouri, 1993 and Zaâfouri et *al.*, 1994a and b and 1995).

The production used is therefore the depigated phytomass of the tree of average biovolume: i.e. the production of the tree of average biovolume divided by the age of the reallocation.

The choice of this type of production estimate, which is somewhat debatable, is imposed by the differences in the age of reallocation in the different ecosystems.

Calculating the biovolume of the average individual requires, as mentioned above, measuring the biovolume of the 15 individuals in each plot. To define the biovolume, it is imperative to measure certain dendrometric parameters (FAO, 1982; Zaâfouri, 1993 and Zaâfouri et *al.*, 1994a and b):

- the maximum and minimum diameters of the crown: measured perpendicular to the ground;
- the height of the crown: this is the height between the lowest green branch and the top of the tree;
- the height of the crown at maximum diameter: this is the height between the lowest green branch and the maximum diameter of the crown.

This biovolume is the volume in space occupied by branches, twigs and leaves. It is, in reality, **a volume of encumbrance**.

The calculation of biovolume refers to simple geometric shapes to which the individuals are related (Barbéro, 1981; Baudin, 1985; Floret, 1988; Zaâfouri, 1993 and Zaâfouri et *al.*, 1994a and b). The geometric shapes found in *A. saligna* in arid and desert areas of Tunisia (Cf. Zaâfouri, 1993) are :

- a cone whose base is either at the top or bottom of the individual;
- two superimposed cones (their bases are in the middle of the individual);
- a sphere ;

-a half-sphere.

It should be noted that these last two forms can be located either at ground level or a few centimetres to 1 m above ground level.

By applying mathematical formulae to these geometric shapes (Cf. Zaâfouri, 1993), we obtain the biovolume of each individual.

The individual with the average biovolume in each plot was then harvested and its dehaired phytomass was weighed in the field. A 1 kg sample was then dried in an oven.

This approach requires a perfect correlation between the dendrometric parameters measured (total height, trunk diameter and crown diameter), the biovolume and the epigeous phytomass. The correlation between these parameters, verified by ourselves (Zaâfouri, 1993), is statistically highly (**) to very highly (***) significant at the 5% probability threshold (Tab. 3).

Table 3. Correlation matrix between dendrometric parameters, biovolume and Epigeous phytomass of *Acacia saligna* (from Zaâfouri, 1993)

	Total height	Trunk	Tree crown	Biovolume

		diameter	diameter	
Total height	-	-	-	-
Trunk diameter	**0,82*** **	-	-	-
Tree crown diameter	**0,78** **	**0,87*** **	-	-
Biovolume	**0,77** **	**0,94*** **	**0,88*** **	-

2.3 METHOD OF STATISTICAL ANALYSIS

The question posed by this study is to explain the influence of environmental factors in arid and desert zones on the survival (success rate) and production (phytomass) parameters of *Acacia saligna.*

Three types of statistical analysis can be used to highlight the link between environmental factors and species parameters (Lamotte, 1962; Decourt et *al.*, 1969; Dagnelie, 1975; Le Tacon, 1975): **(i)** analysis of variance, **(ii)** multivariate analysis based on the correlation coefficient matrix and **(iii)** non-linear analysis. We will not go into the details of these methods, their advantages and disadvantages, which have been extensively described and discussed by the authors cited above and many others.

In this study, we opted for the analysis of variance method followed by an analysis using the Newman-Keuls test.

Analysis of variance is used to determine whether or not a set of samples is homogeneous according to quantitative criteria, and the Newman-Keuls test is applied to simple random samples and to stands of unequal size (Dagnelie, 1975). This is the case for this study.

The approach applied in this analysis can be summarised as follows: when the analysis of variance rejects the hypothesis of equality of a parameter (survival rate or production) in the ecosystems of an ecological region or bioclimate type, we determine the probability and significance of this difference. The Newman-Keuls test at alpha 5% determines the ecosystems that differ significantly, the degree of significance and classifies them in descending order into homogeneous groups (Cf. Chapter 3. Tables 11 to 15).

Chapter 2

ECOLOGICAL CONDITIONS OF TUNISIAN ARID AND DESERT ZONES AND REAFFECTED ECOSYSTEMS

1. ECOLOGICAL CONDITIONS IN ARID AND DESERT ZONES: SUMMARY 1.1. CLIMATE AND BIOCLIMATE

We owe this climatic overview of Tunisian arid and desert zones to the work of Coque (1962), Le Houérou (1959, 1969), Floret and Pontanier (1982 and 1984), Saghiri (1991) and Zaâfouri (2020 and 2023).

In terms of rainfall, Tunisia's arid and desert zones lie between the rainfall isohyets of 100 mm/year and 200 mm/year (see Fig. 3). It forms part of the arid zone with a rainfall of between 100 mm/year and 350 mm/year, which covers around 96,000 km^2 , i.e. 3/5ème of the country's total surface area (164,000 km2). Tunisia's arid and desert zones cover around 30,000 km2. This represents 1/5ème of the country's surface area and 1/3 of the surface area of the arid zone.

From a bioclimatic point of view, the arid and desert areas of Tunisia are located in the Mediterranean isoclimatic zone, as defined by Daget (1977 a). They belong to the area of extension of the Mediterranean bioclimate, as defined by Emberger (1955). In 1959, Le Houérou defined two bioclimatic stages and two sub-stages in these areas:

- arid understorey with temperate to mild variation ;
- Saharan upper sub-storey with temperate to hot variation .

In the rest of the text, they will be referred to successively as the lower arid Mediterranean bioclimate and the upper Saharan Mediterranean bioclimate.

The upper understorey arid stage with a cool to temperate variation **(upper arid Mediterranean bioclimate)** is very poorly represented. It appears on the heights of the low mountain ranges (600 mm to 800 m) of Bou Hedma, Orbata and the Matmata range.

The main climatic features of this arid and desert zone are the extreme irregularity of rainfall in space and time, and climatic aridity. The general features of the climate in these areas are :

- low rainfall: annual average of 148 + 100 mm.
- the average annual rainfall (50% or even 100%) could fall in the space of 24 hours;
- a very limited number of rainy days: 15 days/year to 40 days/year;
- Intense thundery rainfall that can exceed 100 mm/h for 5 minutes;
- a rainfall regime, as defined by Musset (1934), of the Winter-Autumn-Spring-Summer (AHPE) or Autumn-Winter-Spring-Summer (HAPE) type. The cold season (Winter or Autumn) is the wettest and the hot season (Summer) is dry;
- an average annual temperature of 20°C and absolute maximums of 45 to 50°C ;
- an annual temperature range of 24°C to 34°C ;
- a high diurnal monthly temperature range of 25°C to 35°C;
- high average potential evapotranspiration of between 1000 mm/year and 1800 mm/year;
- an annual water deficit of 5 times in a wet year (P > 200 mm) and 36 times in a dry year (P < 50 mm);
- a dry period, as defined by Bagnouls and Gaussen (1953), which is very long (9 months/year to 12 months/year) and begins at the end of the cold, wet period (February);
- violent winds from 50 days/year to 80 days/year);
- hot, dry winds (siroccos) from the Sahara from 25 days/year to 75 days/year.

1.2. GEOMORPHOLOGY, GEOLOGY AND SOILS

The authors cited above have the merit of characterising Tunisia's arid and desert zones from the point of view of geomorphological formations (jbels, glacis and plains), geology (limestone, gypsum) and the physico-chemical characteristics of the soils.

This work has defined three types of soil formations with the following general characteristics:

- limestone and/or gypsum crusts and crusts. These formations date from the Quaternary period and fossilise erosion glacis at the foot of mountain ranges and saline soils in depressions (Chotts and Sebkhas);
- Recent sediments evolving towards sub-desert brown soils with a horizon of deep limestone or calcareous-gypsum nodules;
- raw desert climatic soils and sandy accumulations covering a large part of the pre-Saharan zone, essentially the ecological regions of J'fara and Ouara.

The main soil types in these areas, according to Floret and Pontanier (1982), are shown in **table 4**. These soils are either limestone (25%), gypsum (24%) or halomorphic (19%). These types of soil make up around 70% of the soil formations that are biologically unsuitable for agricultural production. Potentially productive soils account for only 25%, and are used for food crops (market gardening and arboriculture). Since the end of the 90s of the XXème century, even limestone, gypsum and halomorphic soils have been used for food crops.

Table 4. Main soil types in Tunisian arid and desert zones According to Floret and Pontanier (1982)

Type of soil	Occupation	
	Surface area (ha)	**Rate (%)**
Soils with sandy formations	570 000	19
Silty soils of glacis and Segui lands and soils of deflation, spreading and lowland areas	390 000	13
Gypsum soils	720 000	24
Skeletal limestone soils: glacis with limestone crust and crusts and mountain slopes	750 000	25
Soils of halomorphic formations: chotts, sebkhas and their borders	570 000	19
Soils of oases and irrigated areas	17 000	-
Total	**3 017 000**	**100**

According to the same authors, the main edaphic and hydric characteristics of soils are :

- low organic matter content in all horizons: less than 2%;
- high levels of limestone or gypsum, sometimes in excess of 30%;
- fine to coarse textures ;
- sometimes high levels of salts (7 mS/cm to 15 mS/cm): sodium chloride, calcium carbonates and sulphates, potassium and magnesium;
- a tendency for the surface of the soil to be covered by a film of capping;
- susceptibility to water and wind erosion due to the structure of the soil and the lack of sufficient plant cover;
- high rainwater runoff and consequently low infiltration in skeletal soils, limestone and gypsum soils, sloping soils or soils covered by a layer of clay;

- The loamy soils are difficult to moisten because of their loose nature. This type of soil has good water retention capacity;
- good infiltration of rainwater in deep sandy soils, but low water storage capacity due to their permeability.

The physico-chemical characteristics of the soils and the lack of plant cover are not favourable factors for water recharge. They do not allow water reserves to be carried over from one year to the next, except in exceptionally wet years and in deep soils with a sandy to sandy-loam texture (Bourges et *al.*, 1975).

1.3. VEGETATION

The characterisation of the vegetation of Tunisian arid and desert zones is a synthesis of the work of : Le Houérou (1959 and 1969, 1995), Le Floc'h (1979), Floret and Pontanier (1982), Schönenberger (1986 a and b and 1987), Chaieb (1991), El Hamrouni (1992), Akrimi et *al.* (1995), Auclair and Zaâfouri, 1996, Akrimi et *al.* (1996), Zaâfouri (1996, 1998, 2020 and 2023), Ouled Belgacem, Zaâfouri (1997), Ferchichi (1999), Jeder et *al.* (2000).

The species nomenclature is that of the Tunisian floras (Cuénot et *al.*, 1954 and Pottier-Alapetite, 1979 and 1981).

The natural vegetation of these arid and desert areas is mainly made up of steppe formations dominated by champhytes and a relic of grass steppe. A forest-steppe based on *Acacia tortilis* (=Acacia *raddiana*) covering around 19,000 ha is confined to the Bled Talah region. Very degraded, sparse scrub based on *Juniperus phoenicea* ssp. *euphoenicea*, *Rosmarinus officianalis* and isolated individuals of *Rhus tripartitum* and *Periploca laevigata* can be found in inaccessible areas of the Orbata-Bou Hedma and Matmata mountain ranges. A few isolated plants of *Calligonum azel*, *Calligonum comosum* and *Ephedra alata* can be found in the barkans of the Nefzaoua, Jérid and the Great Eastern Erg.

The main steppe formations in Tunisia's arid and desert zones are :

-Steppe with *Stipa tenacissima*: this esparto steppe characterises the natural ecosystems of the lower arid Mediterranean bioclimate. It is found in the ecological regions of the Southern Low Plains and the Matmata. A very degraded remnant of this steppe can be found in the J'fara and to the north of the Ouara in the upper Saharan Mediterranean bioclimate.

-Steppe with *Arthrophytum scoparium*: found from the lower arid Mediterranean bioclimate to the upper Saharan Mediterranean bioclimate in the same Alfa area (degradation stage of the Alfa steppe). This steppe colonises glacis, crusts and limestone crusts.

-Steppe with *Arthrophytum schmittianum*: the bioclimatic environment of this steppe is the upper Saharan Mediterranean bioclimate. It occurs in patches in the lower arid Mediterranean bioclimate.

-Steppe with *Artemisia herba-alba*: mainly found in the Southern Low Plains and Matmata regions in the lower arid Mediterranean bioclimate. A few pockets are found in the upper Saharan Mediterranean bioclimate in the ecosystems with alluvial accumulations of the J'fara and Ouara rivers. This steppe results from the degradation of the Alfa steppe on loamy and sandy loam soils.

-Steppe with *Rantherium suaveolens*: characteristic of sandy soils, this steppe is found mainly in the J'fara region in the lower arid Mediterranean bioclimate and in the north of the Ouara region in the upper Saharan Mediterranean bioclimate. It is little represented in the Southern Low Plains and the Matmata.

-Steppe with *Anthyllis sericea*: characteristic of the upper Saharan Mediterranean bioclimate; it occupies limestone and sometimes gypsum crusts. It is found to the south of the J'fara region and in the Ouara region.

-Steppe with *Aristida pungens*: of anthropogenic origin, it can be found in all the arid and desert areas of Tunisia. It occupies sandy soils, dune accumulations and along temporary or episodic watercourses (wadis).

-Azonal steppes: these are steppes linked mainly to the edaphic characteristics of the ecosystem. They are found in all the arid and desert areas of Tunisia. These steppes are made up of halophilous species, *Arthrocnemum indicum, Halocnemum strobilaceum, Suaeda* ssp. *etc.*, in saline environments and calcarophilous or gypsophilous species on limestone or gypsum soils (*Gymnocarpos decander, Atractylis serratuloides, Helianthemum kahiricum, Zygophyllum album, Annarrhinum brevifolium etc.).*

In addition to quantitative decline (surface area), the steppes of Tunisia's arid and desert zones are characterised *by qualitative decline (low plant cover and floristic diversity, disappearance of species of economic value and rarefaction of keystone species). The quantitative and qualitative degradation of the natural vegetation in these areas is a consequence of* the *intensity of the anthropozoic actions carried out over a very long period (overgrazing and clearing) for* the extension of subsistence agriculture. *The status of these steppes as threatened with extinction,* as defined by the UCIN *(1997 and 1998), is very advanced. They are classified as "***Endangered***", or even "***Extinct***" for certain steppes and certain species.*

1.4. ANTHROPOZOIC CONTEXT

The Office du Développement du Sud (ODS, 2021) estimated the human population of the arid and desert areas of Tunisia (the Governorates of Gafsa, Gabès, Médenine and Tataouine) at **1,184,193** in 2014 (Tab. 5).

Table 5. Anthropozoic context of the ecological regions of the areas studied

According to the Office de développement du Sud (ODS) (2021)

	Southern Lowlands (Gafsa)	**Matmata (Gabès)**	**J'fara (Médenine)**	**Ouara (Tataouine)**	**Total**
Population	**250 000**	**407 500**	**377 240**	**149 453**	**1 184 193**
Surface area (ha)	**780 775**	**716 900**	**916 900**	**3 888 920**	**6 302 895**
-Arable land	248 200	24 360	229 523	200 000	702 083
-Course	226 000	414 695	605 843	1 508 579	2 755 117
Forests	206 573	90 561	81 148	-	378 282
-Waste land	99 902	186 984	186	2 180 321	2 467 413
Livestock (head)	**364 504**	**653 339**	**601 025**	**405 670**	**2 204 538**
-Cattle	13 500	7 600	742	342	22 184
-Sheep	282 582	452 000	416 624	314 015	1 265 221
-Goats	65 772	192 200	166 732	79 020	453 724
-Camels	2 650	1 539	16 927	12 293	33 409

The rural population, representing around 80%, is divided into three categories according to the use of natural resources (Zaafouri, 1993; Auclair and Zaâfouri, 1996; Zaafouri et *al.*, 1997; Visser et *al.*, 1997; Ouled Belgacem and Zaâfouri, 1997):

- **Nomadic pastoralists:** these are breeders of sheep, goats and camels. Depending on the rainfall of the year, this category moves around the area of their ecological region in search of grazing land.

- **Farmer-breeders:** these are both land cultivators (arboriculture, cereal growing and sometimes even market gardening) and semi-extensive livestock breeders: sheep, goats and sometimes cattle. The movement of livestock is limited in space.
- **Farmers:** these are farmers who cultivate the land with occasional cereals, dry arboriculture (olive trees) or irrigated crops (market gardening). These populations are becoming increasingly important as the first two are gradually transformed into farmers.

The number of livestock in the ecological regions studied (Tab. 5) was estimated by the ODS in 2014 (ODS, 2021) at 2,204,538 head. Sheep are the main livestock type (1,265,221 head), followed by goats (453,724 head) and camels (33,409 head). Cattle represent 22,184 head.

Sheep rearing, a traditional practice, predominates in almost all Tunisian arid and desert areas (Tab. 5) and goat rearing is becoming important in degraded and desertified areas characterised by the disappearance of good pastoral species and the multiplication of degrading species of little interest to sheep. Camels are being reared in arid areas (Sebkhas, Chotts, saline soils, gypsum and limestone soils). Cattle are raised in stabilisation and fed on fodder and concentrates, and sometimes on spontaneous species exploited in natural ecosystems.

Although the surface area of land in Tunisia's arid and desert zones considered as natural grazing land (rangelands, forests and uncultivated land "Tab. 5") is significant (around 5.27 million hectares), most of it is desert (50% to 75%, successively, in the Governorates of Médenine and Tataouine) or desert land in the south of the Governorates of Gafsa and Gabès. Only 2.7 million hectares are actually grazing land (Zaâfouri *al.*, 1994b). These rangelands are very degraded and very poor in pastoral species (Floret et *al.*, 1973; Schönenberger, 1986b and 1987; Chaieb, 1991; Zaâfouri et *al.*, 1994b; Ferchichi, 1999; Chaieb and Zaâfouri, 2000 and Zaâfouri, 1998, 2020 and 2023).

The natural ecosystems of the arid and desert zones, which used to play a dominant role in the pastoral farming system, are gradually giving way to food crops, mainly arboriculture (olive trees) and market gardening. The intensification of food crops through state subsidies has reduced the space available to these pastoral ecosystems, pushing them back to marginal lands (glacis, regs and sandy soils). Moreover, these plant groups, which are overexploited and have a very high "Endangered" or even "Extinct" status, continue to suffer from pressure from livestock.

These natural ecosystems, which make up the rangelands of the arid and desert zones (around 2.7 million hectares) only manage to produce $0.18.10^8$ UF/year (at a rate of 30 UF/ha/year). They produce only 15% of livestock requirements, estimated at $5.28.10^8$ UF/year (Zaâfouri et *al.*, 1994b and Ferchichi, 1999). The fodder deficit in arid and desert areas is therefore 75%.

To deal with the degradation of natural rangelands and make up for the fodder deficit, national policy has resorted to reallocating these ecosystems by introducing non-native shrubs. The aim of these introductions is to safeguard the environment of the rangelands, improve their fodder production by creating standing fodder reserves and activate the vegetation dynamics of these ecosystems through natural regeneration.

2. ECOLOGICAL CONDITIONS OF REAFFECTED ECOSYSTEMS

It should be remembered that the zoning of the Tunisian arid and desert areas has made it possible to retain only 4 ecological regions (Basses Plaines Méridionales, Matmata, J'fara and l'Ouara).

2.1. CLIMATE DURING THE 1980-1990 DECADE

From a climatic point of view, we feel it would be useful to point out that our analysis of climatic conditions in these regions only covered the decade 1980-1990:

2.1.1. The reallocation of natural ecosystems in these areas took place at the end of the 1970s;

2.1.2. The soil's water reserves do not carry over from one year to the next, except in rainy years, which only occur one year in five, and in deep soils with a balanced texture. However, these types of soil are not very extensive and are reserved for agricultural production.

The absence of weather stations in each ecological sector or ecosystem has led us to limit ourselves to climatic data from the main stations: Gafsa for the Southern Low Plains, Gabès for the Matmata, Médenine for the J'fra and R'mada for the Ouara. Although it is tricky to generalise climate data from a regional weather station to all ecosystems, there is no other way to do so. However, there is no other way of characterising the climate on the scale of an ecosystem or even an ecological sector.

2.1.3. Rainfall

The rainfall characteristics of the ecological regions of the Tunisian pre-Saharan zone during the decade 1980-1990 are shown in Tables 6 and 7. The main climatic characteristics of this decade are :

-Average annual rainfall of :

* 200.3 mm in the J'fara region (Médenine station) ;

* 197.1 mm in the Matmata region (Gabès station),

* 187.5 mm in the Basses Plaines Méridionales region (Gafsa station) ;

* 95.2 mm in the Ouara region (R'mada station).

-Dry to moderately dry years (P < 150 mm) vary from 5 to 7 years depending on the region. Moderately wet years (150 mm to 200 mm) are 4-year years.

-Exceptionally wet years (P > 200 mm) only occurred in one year out of five.

-The rainy season coincides with the cold season (November-February) and the dry season begins just at the end of the wet and cold period (February-March), coinciding with high temperatures.

Table 6. Climatic characteristics of ecological regions in the decade 1980-1990: Monthly average

Ecological region	Reference station	Climatic parameters	Jan.	Feb.	March	April	May	June	Jul.	August	Seven.	Oct.	Nov.	Dec.	Means.
Southern Lowlands	Gafsa	Maximum temperature (° C)	14,8	17,4	20,6	24,7	29,1	34,3	37,5	36,8	32,6	26,9	20,4	15,9	**25,9**
		Minimum temperature (° C)	3,8	5,2	7,9	11,1	15,1	19,3	21,9	21,9	19,2	15,0	9,2	5,0	**12,9**
		Average temperature (° C)	10,7	H,4	14,4	17,8	22,0	26,5	28,9	28,7	25,8	21,1	15,1	11,0	**19,4**
		Rainfall (mm)	41,4	7,7	17,0	15,7	11,1	8,3	EO	8,7	18,6	20,1	20,7	19,3	**189,6**
		Evaporation Piche (mm)	101,3	115,1	169,0	215,0	334,6	334,6	375,6	278,4	235,0	159,9	112,3	87,0	**2183,2**

Matmata	Gabès	Maximum temperature (° C)	16,5	16,5	18,4	22,3	25,0	27,9	31,1	31,8	30,2	26,9	22,1	17,7	**23,8**
		Minimum temperature (° C)	7,6	8,6	10,8	14,1	17,4	21,2	23,2	24,2	22,3	18,7	13,3	8,9	**15,8**
		Average temperature (° C)	12,0	13,5	15,4	18,2	21,2	24,5	27,2	28,0	26,2	22,8	17,7	13,3	**19,8**
		Rainfall (mm)	30,2	12,3	15,5	13,0	4,4	3,6	0,7	0,9	8,2	37,4	34,4	36,5	**197,1**
		Evaporation Piche (mm)	158,8	164,4	169,2	162,3	184,5	176,7	200,5	194,2	178,9	156,0	157,5	146,0	**2049,0**
J'fara	Medenine	Maximum temperature (° C)	16,9	19,5	21,5	25,8	29,1	33,2	36,2	36,5	33,4	29,0	22,8	16,3	**26,7**
		Minimum temperature (° C)	7,0	8,3	10,2	13,9	16,5	19,9	22,3	23,2	21,4	18,3	12,6	8,5	**15,2**
		Average temperature (° C)	H,9	13,9	15,8	19,8	22,8	26,6	29,3	29,8	27,4	23,6	17,7	13,3	**21,0**
		Rainfall (mm)	26,5	12,2	10,8	9,3	4,1	3,0	1,5	1,2	7,0	32,3	50,0	42,4	**200,3**
		Evaporation Piche (mm)	110,6	119,4	156,8	180,1	222,1	224,6	245,9	209,4	160,5	131,8	119,0	102,0	**1982,2**
Ouara	R'mada	Maximum temperature (° C)	15,8	18,9	21,5	26,7	30,9	35,0	37,3	37,4	34,1	28,9	22,1	16,5	**27,1**
		Minimum temperature (° C)	6,5	10,0	13,3	17,0	17,0	19,8	21,6	22,4	20,6	17,1	12,0	7,3	**15,4**
		Average temperature (° C)	H,1	13,5	15,8	20,0	24,0	27,4	29,4	29,9	27,3	23,0	17,0	11,9	**20,9**
		Rainfall (mm)	15,3	6,2	15,0	4,8	3,6	3,1	1,0	1,6	3,2	2,2	18,1	21,1	**95,2**
		Evaporation Piche (mm)	133,4	163,1	195,0	248,9	306,9	315,9	329,9	318,1	248,7	197,2	158,8	121,0	**2736,9**

Table 7. Climatic characteristics of ecological regions :
(1) Annual average for the decade 1980-1990
(2) Average of 70 years

Ecological region	**Reference station**	**Average annual rainfall (mm)**	**Temperatures (in⁰ C)** Maxi (M)	**Mini (m)**	**Thermal amplitude (M-m)**	**Average**	**Evaporation Piche (mm)**	**Emberger rainfall-thermometer (Q)₂**	**Dry period** Bagnouls and Gaussen (P < 2T)	**Aubréville (P < 30 mm)**
Low Plains Méridionale	Gafsa **(1)**	189.6	37.5	3.8	33.7	19.4	2183.2	17.6	11	11
	(2)	163.0	38.1	3.9	34.2	19.3	2858.0	15.8	12	12

Matmata	Gabès									
	(1)	197.1	31.8	7.6	24.2	19.8	2049.0	10	10	8
	(2)	187.0	32.4	5.9	26.8	19.3	2022.0	12	12	10
J'fara	Medenine									
	(1)	200.3	36.5	7.0	29.5	21.0	1982.2	21.1	9	9
	(2)	144.0	36.8	6.2	30.6	20.5	1096.0	16.0	12	12
Ouara	R'mada **(1)**	95.2	37.4	6.5	30.9	20.9	2736.9	9.4	12	12

2.1.4. Temperatures

Temperatures are less subject than rainfall to large fluctuations within the same ecological region. Temperature variations in the 4 ecological regions are small and very close to each other. Moreover, they do not differ significantly from the average over a 70-year period.

Temperatures in the 1980-1990 reassignment decade (Tab. 6 and 7) are characterised by :

- average annual temperatures of 19.5°C to 20.9°C. They are 19.5°C, 20°C, 20.8°C and 20.9°C, respectively, in the Southern Low Plains, Matmata, J'fara and Ouara;
- very high average highs for the hottest month (M): 37.5°C in the Southern Low Plains; 31.8°C in the Matmata; 36.5°C in the J'fara and 37.4 in the Ouara ecological region;
- The average minimum temperatures for the coldest month (m) ranged from 3.8°C to 7.6°C. The lowest is 3.8°C (Basses Plaines Méridionales region) and the highest is recorded in the Ouara (6.5°C), J'fara (7°C) and Matmata (7.6°C) regions.
- An annual temperature range (M - m) of 24.2°C to 33.7°C. The highest is recorded in the Southern Low Plains (33.7°C) and the lowest in the Matmata (24.2°C). In the Ouara and J'fra regions, the amplitudes were 30.9°C and 29.5°C respectively.
- Absolute temperatures in the hot season can reach 40°C to 45°C, and even 50°C in the Ouara region (R'mada station).

2.1.5. Evaporation and evaporative demand

Average annual Piche evaporation is very high in all ecological regions. It varies from 1979.2 mm/year to 27218 mm/year (Tab. 6 and 7). Under the influence of the Mediterranean, the J'fara and Matmata regions are characterised by the lowest levels of evaporation (1979.2 mm/year and 2040.7 mm/year respectively). On the other hand, high levels of evaporation are recorded in the continental ecological regions: 2721.8 mm/year in Ouara and 2471.2 mm/year in the Southern Low Plains.

Annual evaporative demand in these regions is very high compared with annual rainfall contributions. It varies from 10.5 times in the wet season (September-March) to 31.5 times in the dry, hot season (April-August). Monthly evaporation in relation to monthly rainfall is 2.4 times to 26.3 times in wet periods and 12.4 times to 375.6 times in dry and hot periods.

2.1.6. Rainfall regime and length of dry period

According to Le Houérou (1959 and 1969) and Floret and Pontanier (1982), the arid and desert zones of Tunisia are characterised by two types of rainfall regime as defined by Musset (1934):

-A **winter-autumn-spring-summer** (**WASS**) rainfall pattern;

-an **Autumn-Winter-Spring-Summer** (**AHPE**) plan.

The cold seasons are therefore the rainy ones and the hot ones are dry.

These rainfall patterns, recorded over a 70-year rainfall series (authors above), were evident during the 1980-1990 decade in the ecological regions studied. They are confirmed by Zaâfouri (2020 and 2023) for a period of 20 years in the Bled Talah region located at the

northern limit of arid and desert zones.
The continental ecological regions (Basses Plaines Méridionales and Ouara) are identified by a **HAPE-type** rainfall regime, with Winter being the wettest season (Fig. 4). As for the two regions that open onto the Mediterranean (Matmata and J'fara), they are characterised by an **AHPE-type** rainfall pattern, with Autumn being the wettest season (Fig. 4). Moreover, autumn and winter in the J'fara region are characterised by almost the same amount of rain (89.3 and 81.1 mm respectively).
The duration of the dry period (Fig. 5), as defined by Bagnouls and Gaussen (1953), during the 1980-1990 decade was 11 to 12 months/year in the continental ecological regions (Basses Plaines Méridionales and Ouara) and 9 to 10 months/year in those opening onto the Mediterranean (Matmata and J'fara).

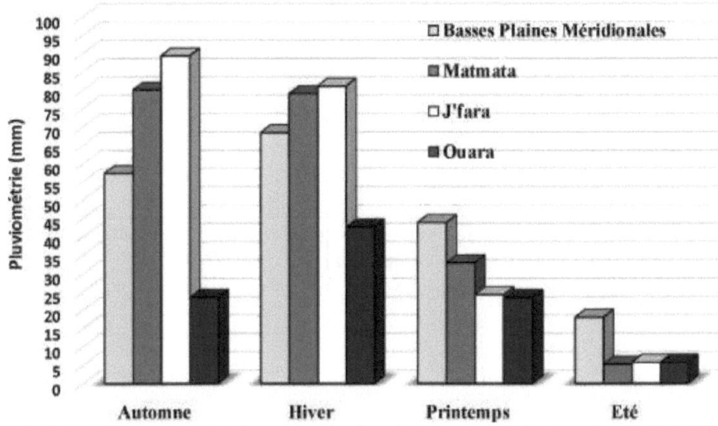

Figure 4: Rainfall patterns in the ecological regions studied during the 1980-1990 decade

These durations for a decade are almost the same for a 70-year rainfall series calculated by Le Houérou (1959 and 1969) and Floret and Pontanier (1982) and Zaafouri (2020 and 2023) for a 20-year period (9 months/year to 10 months/year). The dry period begins just after the rainy season and extends from February or March to October or November. A dry period of 12 months/year in these arid and desert areas is quite common (authors quoted above).

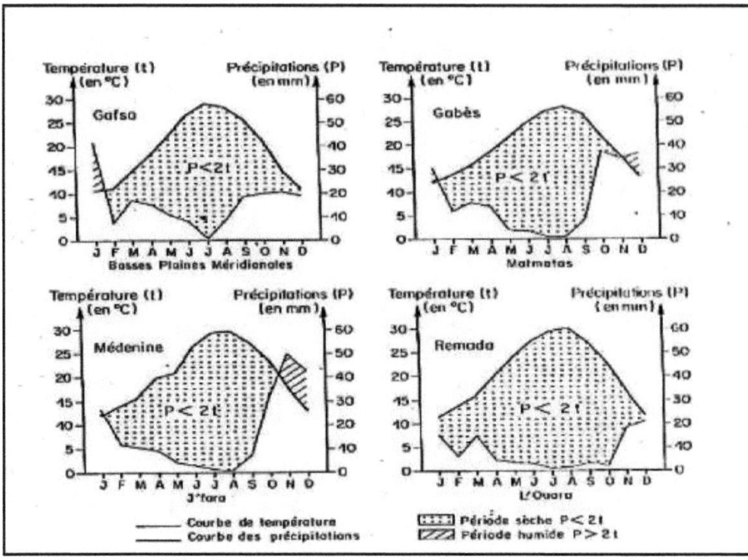

Figure 5. Duration of the dry period in the ecological regions studied during the decade 1980-1990

2.2. GEOMORPHOLOGY AND SUBSTRATE (Tab. 8 to 11)

2.2.1. Geomorphology (Fig. 6)

The reallocated ecosystems are located on 4 types of geomorphological units: Glacis, Plains, Depressions and Wadi beds. Around ¾ (70%) of these ecosystems are located on glacis. Plains ecosystems occupy only about $1/5^{ème}$ (17%).

Reallocated ecosystems located on glacis dominate in the Matmata, J'fara and Ouara ecological regions (60% to 100%). On the other hand, around 60% of the reaffected ecosystems of the Southern Low Plains are located on plains.

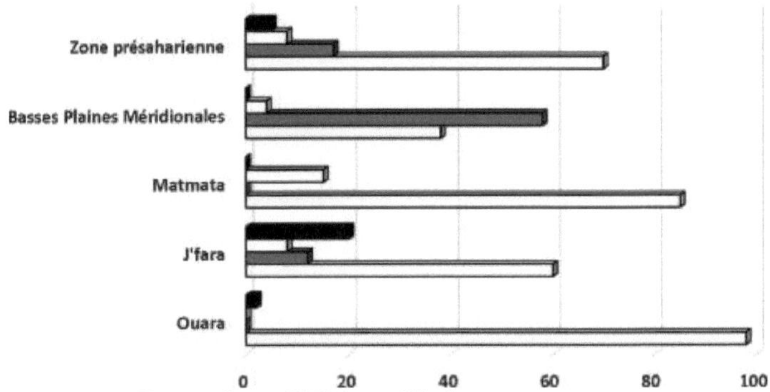

■ Wadi beds□ Depressions UPlains □ Glacis

Figure 6. Frequency of geomorphological units in reallocated ecosystems (% of surface area)

2.2.2. Substrates (Fig. 7)

30% to 50% of the surface area of the reallocated ecosystems is located on limestone or gypsum substrates in the form of crusts or encrustations. The limestone substrate dates from the Middle and Early Quaternary (Villafranchian for the oldest) and the gypsum substrate dates from the Recent Quaternary. Soils without substrate, referred to in this study as "soils without bedrock", represent only 1/5th (20%) of the surface area of the reallocated ecosystems.

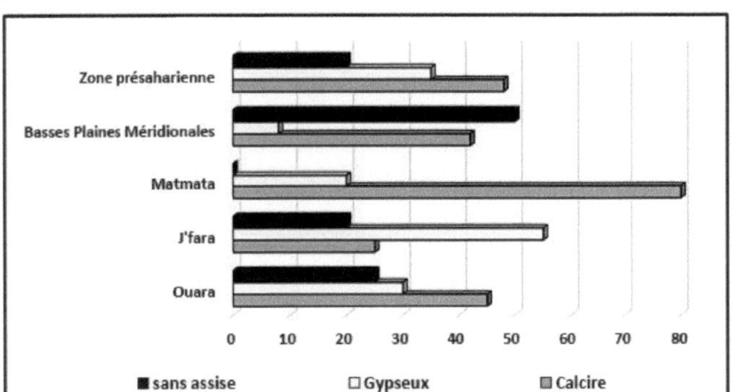

Figure 7. Frequency of substrate types in reallocated ecosystems (% of surface area)

Table 8. Geomorphological, edaphic and floristic characteristics of the reallocated

sectors and ecosystems of the Southern Lowlands region

Ecological sectors	Ecological systems (ecosystems)		Symbol	Furniture layer thickness (cm)	Water regime	Soft layer content (%)				Salinity soft layer (mS)
	Type of soil	Vegetation facies Degree of disturbance				Organic matter	Gypsum	Total limestone	Active limestone	
Covered **glacis** of the ancient Villafranclrienne surface with salmon-coloured **limestone** crust. Degradation **steppe** of *Juniperus phoenicea* ssp. *Sviphoenicea*	Slightly developed site with a thick sandy loose layer on a limestone crust	*Stipa tenacissima, Helianthemum Iippii ssp. Sessiliflorum* and *Salvia verbanica.* **Gradient**	**GCl**	60 to IOO	I = P	0,04 à 0,7	0	12,6 à 35	3 à 10	0,5 à 1,8
	Slightly developed site with a very thick loose sandy-loam layer with no identified substrate	*Arthrophytum Schmittianum* and *Arthrophytum scoparium.* **Moderately degraded**	**GC2**	> IOO	I = P	0,3 à 0,4	0	15 à 32,2	3 à 8	0,6 à 2,2
Gypsum glacis with no soil cover. **Steppe** with *Zygophyllum album* and *Annarrhinum brevifolium*	Erosion channel with gypsum crust and encrustation	*Annarrhinum brevifolium, Helianthemum ellipticum* and *Gymnocarpos decander.* **Very degraded**	**GGl**	0	I = P-R	0,2 à 0,4	23,2 à 55,0	12,7 à 21,8	12 à 20	2,6 à 3,4
Sandy plains. Steppe with *Arthrophytum Schmittianum, Artemisia herb a-alb a* and *Hedysarum carnosum*	Slightly developed site with a very thick loose sandy-loam layer with an eolian veil, with no identified substratum	*Helianthemum Iippii ssp. Sessiliflorum QtAristida pungens.* **Moderately degraded**	**Pl**	> 100	I = P	0,1 à 1,7	0	13,8 à 52,5	0 à 15	0,4 à 1,7
	Slightly evolved	*Asphodelus tenuifolius*	**P2**	> 100	I = P	0,04 à 0,3	0	16,1 à 29,9	1 à 4	0,4 à 0,5

	site with very thick loose sandy layer to meso-nebkas, with no identified substrate	and *Aristida pungens*. **Moderately degraded**								
Depression formed in the Mio-Pliocene. Sub-azonal **steppe** to *Ziziphus lotus* and *Nitraria retusa*	Slightly developed site with a thick loose sandy loam layer with no identified substrate	*Ziziphus lotus* and *Salvia verbanica*. **Moderately degraded**	**Dl**	60 à 100	I = P+R	0,6 à 0,9	0,2 à 1	23,6 à 28,2	1 à 4	0,5 à 1,2

I: Infiltration

P: Rain

R: Runoff

Table 9. Edaphic and floristic characteristics of the sectors and ecological systems reallocated in the Matmata region

Ecological sectors	**Ecological systems**		**Symbol**	**Furniture layer thickness (cm)**	**Water regime**	**Soft layer content (%)**				**Salinity soft layer (mS)**
	Type of soil	**Vegetation facies Degree of disturbance**				**Organic matter**	**Gypsum**	**Limestone Total**	**Active limestone**	
Covered **glacis** of the ancient Villafranclrienne surface with salmon-coloured **limestone** crust **Steppe** to *Arthrophytum Schmittianum* and *Gymnocarpos decander*	Erosion site with a very thin loose sandy layer on limestone crust reworked by subsoiling	*Thymelaea microphylla* and *Atractylis Serratuloides*. **Gradient**	**GC3**	O to 20	I = P	0.5	0.4	42.5	3	1.4
	Slightly developed site with a loose, shallow	*Thymelaea microphylla* and *Polygonum equi setiforme*.	**GC4**	20 à 60	I = P + R	0.2	0.2 à 0.6	15.9	3 à 11	1.1 à 1.4

	sandy layer on a limestone crust reworked by subsoiling and ESC work (tabias)	**Gradient**								
	Erosion site with a very thin loose sandy-loam layer on a limestone crust	*Arthrophytum Scoparium* and *Atractylis Serratuloides* - **Gradient**	**GC5**	0 à 20	I = P-R	0.5	0.3	21.3	4.5	2.4

Table 10. Edaphic and floristic characteristics of the sectors and ecological systems reallocated in the J'fara region

Ecological sectors	Ecological systems		Symbol	Furniture layer thickness (cm)	Water regime	Soft layer content (%)				Salinity soft layer (mS)
	Type of soil	Vegetation facies Degree of disturbance				Organic matter	Gypsum	Total limestone	Active limestone	
Glacis covered by the old surface Dismantled Villafranclrienne with salmon-coloured **limestone** crust. **Steppe** to *Rantherium Suaveolens* and *Asphodehts*	Slightly developed site with a loose, shallow sandy-loam layer on a limestone crust	*Rantherium Siiaveolens. Artemisia campestris. Thytnelaea microphylla* and *Aristida pungens.* **Gradient**	**GC6**	20 à 60	I = P-R	0.2 à 1	0.03 à 2.8	12.7 à 58.5	1 à 4.5	0.5 à 1.6
	Slightly developed site with a loose, thin sandy layer on a limestone crust with mesonebkhas and a sand sheet	*Rantherium Siiaveolens. Artemisia campestris Thymelaea microphylla* and *Aristida pungens.* **Gradient**	**GC7**	20 à 60	I = P	0 à 0.5	0.1 à 5.3	16.5 à 50.8	0.5 à 15.5	0.6 à 1.6

refr news	Slightly evolved site with thick sandy loose layer on thick limestone crust with meso-nebka and sand sheet	*Rantherium Siiaveolens. Artemisia campestris, Thymelaea microphylla* and *Aristida pungens.* **Gradient**	**GC8**	60 à 100	I = P	0 à 0.5	0.03 à 0.7	9 à 26.6	0 à 5	0.5 à 1
Covered **glacis** with **gypsum** crust and crusts. **Steppe** to *Rantherium Suaveolens* and *Lygeum spar turn*	Slightly developed site with a loose sandy loam layer on a gypsum crust	*Rantherium Suaveolens. Thymelaea microphylla* and *Gymnocarpos decander.* **Very degraded**	**GG2**	20 à 60	I = P	0.05 à 0.3	0.4 à 21.20	11.1 à 25.5	0 à 5.5	0.5 A 3
Sandy plain Steppe with *Arthrophytum Schmittianum* and *Aristida pungens*	Slightly developed site with a loose, thin sandy layer on a limestone crust	*Arthrophytum Schmittianum* and *Aristida pungens.* **Very degraded**	**P3**	20 à 60	I = P	0.1 à 0.4	0à0,2	11.5 à 49	0.5 à 2	0 à 0.6
	Slightly evolved site with a very thick loose sandy layer with no identified meso-nebka substratum and sand sheet	*Arthrophytum Schmittianum* and *Aristida pungens.* **Good condition**	**P4**	> IOO	I = P	0.1 à 0.2	0.04 à 1	11.5 à 16,3	0 à 13	0 à 0.5
	Aeolian-fed Iialomorphic sediment with a loose, thin sandy layer on a gypsum crust	*Pituranthos tortuosas.* **Gradient**	**P5**	20 à 60	I = P	0 à 0.4	10 à 15.7	15,7 à 21.3	1 à 8.5	5,2 à 10.2
Low and medium wadi terraces Steppe with *Rantherium Suaveolens* and *Artemisia Campestris*	Slightly developed alluvial site with a very thick loose sandy layer with no identified substratum	*Artemisia Campestris* and *Gymnocarpos decander.* **Gradient**	**BMTl**	> 100	I = P + R	0.13 à 0.53	0,1 à 3	12.5 à 52.1	0.1 à 12.5	0.4 à 4,8
	Aeolian-fed Iialomorphic sediments with a thick, loose sandy layer on a gypsum crust	*Imperata Cylindriea* and *Arthrocnemum indicum.*	**BMT 2**	60 à 100	I = P + R	Where 1	OàlO	0 à 11.5	0	5 à 19.2

	with a slick layer.	**Good condition**								
Depression with **gypsum** crust and crusts **Steppe** with *Rantherium Suaveolens* and *Lygeum spar turn*	Erosion on gypsum crust	*Artemisia Campestris* and *Aristida pungens.* **Very degraded**	**D2**	0 à 20	I = P-R	0 à 0,3	1.4 à 75.3	15 à 18.5	2 à 4,5	3.1 à 6.7
	Slightly evolved sandy loose layer on limestone to meso-nebka crust	*Artemisia Campestris* and *Aristida pungens.* **Very degraded**	**D3**	20 à 60	I = P + R	0,2 à 0.4	1.4 à 8.5	13 à 17,6	1.5 à 4	2,3 à 3
	lialomorphic sediment with a loose, thin sandy layer on a gypsum crust	*Arthrocnemum indicum* and *Aerolopus littoralis.* **Gradient**	**D4**	20 à 60	I = P + R	0.2 à 0.4	1 à 10.5	15.5 à 35.3	3,5 à 16.5	5 à 11.5

Table 11. Soil and plant characteristics of the sectors and ecological systems reallocated in the IOuara region

Ecological sectors	**Ecological systems**		**Symbol**	**Furniture layer thickness (cm)**	**Water regime**	**Soft layer content (%)**				**Salinity soft layer (mS)**
	Type of soil	**Vegetation facies Degree of disturbance**				**Organic matter**	**Gypsum**	**Total limestone**	**Active limestone**	
Covered **glacis** of the old dismantled Villafranclrienne surface with salmon-coloured **limestone crust** **Steppe** with *Anthyllis sericea* ssp. *Heoniana and Gymnocarpos decander*	Erosion site with a loose, shallow sandy layer on a limestone crust reworked by subsoiling	*Anthyllis sericea* ssp. *Heoniana* and *Arthrophytum Schmittianum*. **Gradient**	**GC9**	20 à 60	I = P	0.2 à 0.4	0.4 à 0.7	18.2 à 35.5	0 à 11.5	0.8 à 1.2
	Slightly developed site with a very thick loose sandy layer with no identifie	*Arthrophytum Schmittianum* and *Helianthemum Iippii var. intricatimi.* **Very degraded**	**GClO**	> 100	I = P	0.1 à 0.3	0.3 à 0.9	15,4 à 18.2	0 à 14,5	0.4 à 1.4

	d substrate									
More or less covered **gypsum glacis Steppe** *with Anthyllis sericea* ssp. *Heoniana and Gymnocarpos decander*	Slightly developed site with a loose sandy-silty layer over a gypsum crust	*Arthrophytum Schmittianum* and *Anthyllis sericea* ssp. *Heoniana*. **Very degraded**	**GG3**	20 à 60	I = P	0.3 à 1.2	19,4 à 29.5	10.2 à 17.5	0 à 7	0,5 à 3.5
	Eolian veil erosion on gypsum crusts	*Arthrophytum Schmittianum* and *Atractylis Serratuloides*. **Very degraded**	**GG4**	0	I = P-R	0.1 à 1.1	55,2 à 69.9	13.6 à 14.4	6 à 12	2,4 à 4.5
Low and medium wadi terraces Sub-azonal **steppe** with *Retama raetam and Nitraria retuşa*	Slightly developed site with a loose, thin sandy-silty layer over nodular crusts	*Retama raetam* and *Polygonum equi setiforme*. **Very degraded**	**BMT3**	20 à 60	I = P+R	0,1 à 0,5	14.5 à 18.2	2,5 à 14.5	1.5 à 4.5	0.2 à 4.5

The limestone substratum is dominant in the Matmata (around 80%), Ouara and Basses Plaines Méridionales (45% and 42% respectively) regions. Gypsum substratum is found mainly in the J'fara (> 55%) and Ouara (30%) ecological regions. Unsubstrate soils are represented in the Basses Planes Méridionale ecological region (50%), absent in the Matmata and poorly represented in the J'fra and Ouara (20% and 25% respectively).

2.3. SOILS (Tab. 8 to 11)

2.3.1. Types, salinity and water behaviour

The soils of the reaffected ecosystems of the Tunisian arid and desert regions are **little evolved, non-climacic soils with aeolian input**. **Mixed input soils** (aeolian, alluvial and colluvial) are poorly represented and are found in certain ecosystems of plains, depressions and wadi beds. **Erosion soils, crude mineral soils and halomorphic soils** are fairly well represented. These types of soil are characterised by very low organic matter content (< 2%) and by an almost generalised film of capping.

The 6 reallocated ecosystems in the Southern Lowlands are characterised by **soils that have evolved little, but not climatically, as a result of mixed inputs (eolian and alluvial)**. Only one ecosystem is of the **gypseous water erosion soil** type. Total limestone content is generally high (13.8% to 59.1%) and active limestone content in some soils is also high (>15%). The pH is slightly basic (7.5 to 8.5). The soils of the ecosystems reallocated in this

region are never or only slightly salty (C.E < 4 mS/cm).
The soils of the ecosystems in this ecological region are fairly good reservoirs of water, with the exception of the gypsum water erosion ecosystem. Soils with a more or less impermeable silty-sandy or silty-clay horizon at depth offer the most favourable water conditions for vegetation. In addition, certain types of ecosystems have soils that have the disadvantage of being filtering (low retention capacity) and others have a crust or impermeable surface crust or a layer of capping that accentuates run-off and reduces infiltration.
The 3 reaffected ecosystems in the Matmata region are defined by **soils that have not evolved much from water erosion**, reworked by subsoiling. They are rich in limestone (15% to 42%). These soils are never salty (C.E < 2.5 mS/cm) and the total limestone content is high (13.8% to 59.1%), with some soil types being rich in active limestone (> 11%). The pH is slightly basic (7.5% to 8.5%). The water balance, favoured by subsoiling and water and soil conservation works (tabias), is generally positive (infiltration = precipitation + run-off).
The 12 reallocated ecosystems in the J'fara region are characterised by soils that are, for the most part, **not very developed due to aeolian or mixed** (aeolian and hydric) **contributions**. Ecosystems with **halomorphic soils** are also represented, but are not very well developed. Depending on the nature of the substrate, the soils of ecosystems with a limestone substrate are rich in total limestone (from 9% to 58.5%) and active limestone (>11%) and those with a gypsum substrate are rich in gypsum, which can be as high as 21%. The pH is neutral to slightly basic (7 to 8.9). Salinity is low (C.E. < 4 mS/cm) in most ecosystems. However, it is high in halomorphic ecosystems (5 mS/cm to 15.6 mS/cm). Rainwater infiltration is low in most ecosystems due to the outcropping of crusts and crusts, and is high in the ecosystems of plains, low and medium terraces and wadi beds.
Reallocation in the Ouara region was carried out in 5 ecosystems. Almost all of the soils were **poorly developed, non-climatic soils with aeolian or alluvial inputs**, or **mixed** soils (aeolian and hydric). Soils **with little development due to water and wind erosion** are also represented, but are infrequent. The pH is slightly basic (7.5 to 8.7). These soils are rich in limestone (10.2% to 35.5%) or gypsum (up to 69.9%) depending on the nature of the substrate. They are never salty (C.E < 2 mS/cm). Depending on the layout of the toposequence, the soil water balance is either negative due to run-off or positive in depressions, low and medium terraces and wadi beds (infiltration + run-off).

2.3.2. Texture and Structure (Fig. 8)

The sandy to sandy-loam texture of the soils is predominant in the reaffected ecosystems of Tunisia's arid and desert zones. The deep horizons (> 80 cm) are frequently characterised by a fine silty texture, sometimes clayey. Wadi bed soils are either alluvial or colluvial. Soils with a sandy-gypsum or sandy-limestone texture are widely represented. The structure of these soils is generally polyhedral.

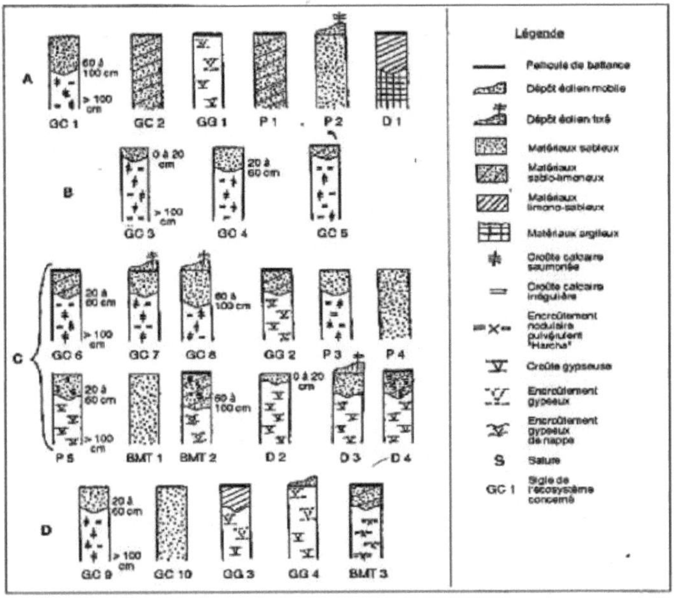

Figure 8. Soil profiles of reallocated ecosystems

A: Southern Low Plains **B:** Matmata **C:** J'fara **D:** l'Ouara

2.3.3. Thickness of loose layer

The thickness of the soft layer in the soils of reallocated ecosystems depends on the position of the ecosystem on the toposequence and therefore the type of ecological sector. 4 soft layer thicknesses are distinguished:

- Very thin loose layer: 0 cm to 20 cm.
- Shallow loose layer: 20 cm to 60 cm.
- Thick loose layer: 60 cm to 100 cm.
- Very thick loose layer: > 100 cm

In general, ecosystems located on glacis are characterised by soils with a very thin to thin loose layer (0 cm to 20 cm and 20 cm to 60 cm). Those located on plains and wadi beds are characterised by soils with a thick to very thick loose layer (60 cm to 100 cm and even more than 100 cm). Almost 1/3 (60%) of the surface area of soils in reallocated ecosystems is defined by a loose layer less than 60 cm thick (very thin to thin).

In the Southern Low Plains (Fig. 9), 84% of the surface area of the reallocated ecosystems is characterised by soils with thick to very thick loose layers (60 cm to +100 cm). In contrast, in the Matmata (Fig. 9), 100% of the reallocated ecosystems have very thin to thin loose layers (< 60 cm). And 60% to 70% of the areas of reaffected ecosystems in J'fara and Ouara (Fig. 9) are identified by very thin to thin loose layers (< 60 cm). Some ecosystems in the latter two regions are characterised by the outcrop of a gypsum crust and those in the Matmata by a limestone crust or crusts.

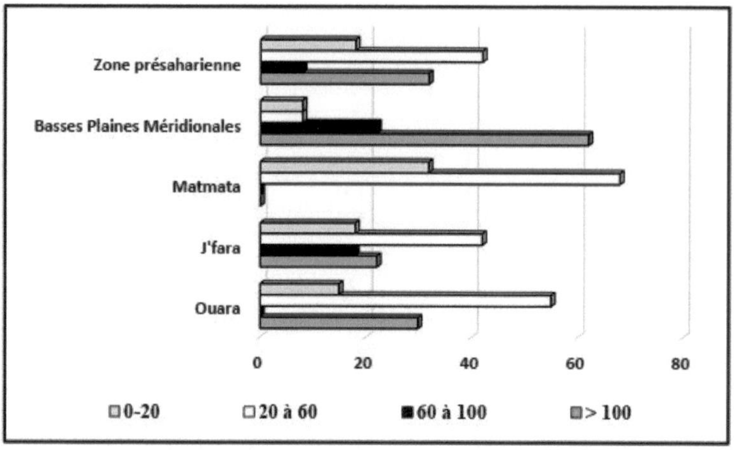

Figure 9. Frequency of the thickness of the soft layer of soils in reallocated ecosystems (% of surface area)

2.4. VEGETABLE GROUPS AND DEGRADATION FACIES (Tab. 8 to 11)

We note that vegetation cover in reaffected ecosystems in arid and desert zones is very low (10% to 30%). It does not exceed 40% in ecosystems where ecological conditions are the most favourable (Zaâfouri, 1993, 1998, 2000 and 2003).

The vegetation of the reallocated ecosystems is broken down into plant groups and degradation facies.

2.4.1. Groups and facies of the Southern Low Plains

The plant groups and facies described in this region are :

* Degradation of the *Juniperus phoenicea* ssp. *euphoenicea* forest. It includes 2 facies:

* *Stipa tenacissima, Helianthemum Lippii* ssp. *sessiliflorum and Salvia verbanica*

* *Arthrophytum schmittianum* and *Arthrophytum scoparium*

* Grouping with *Zygophyllum album* and *Annarrhinum brevifolium*. This consists of a single facies with *Annarrhinum brevifolium*, *Helianthemum ellipticum* and *Gymnocarpos decander*.

* Group with *Arthrophytum schmittianum*, *Artemisia herba-alba* and *Hedysarum carnosum*. It contains 2 facies:

* *Helianthemum Lippii* ssp. *sessiliflorum* and *Aristida pungens*

* *Asphodelus tenuifolius* and *Aristida pungens*

* Sub-azonal group with *Ziziphus lotus* and *Nitraria retusa*. Only one facies is represented: *Ziziphus lotus* and *Salvia verbanica* facies.

2.4.2. Groups and facies in the Matmata

A single grouping covers the reallocated ecosystems of the Matmata:

* Group with *Arthrophytum scoparium* and *Gymnocarpos decander*. This grouping comprises 3 facies:

* *Thymelaea microphylla* and *Atractylis serratuloides*

* *Thymelaea microphylla* and *Polygonum equisetiforme*

* *Arthrophytum scoparium* and *Atractylis serratuloides*

2.4.3. Groups and facies of the J'fara

The plant groups and facies described in the J'fara are :

* Group with *Rantherium suaveolens* and *Asphodelus refractus*. 2 facies are represented:

* *Rantherium suaveolens*, *Artemisia campestris*, *Thymelaea microphylla* and *Aristida pungens*

* *Rantherium suaveolens*, *Artemisia campestris* and *Aristida pungens*, covering two different ecosystems

-Group with *Rantherium suaveolens* and *Lygeum spartum*. It consists of a single facies with *Rantherium suaveolens*, *Thymelaea microphylla* and *Gymnocarpos decander*.

-Group with *Arthrophytum schmittianum* and *Aristida pungens*. It contains 2 facies :

*Arthrophytum *schmittianum* and *Aristida pungens*

*Pituranthos *tortuosus*

-Group with *Rantherium suaveolens* and *Artemisia campestris*. It includes 2 facies:

*Artemisia *campestris* and *Gymnocarpos decander*

*Imperata *cylindrica* and *Arthrocnemum indicum*

-Group with *Rantherium suaveolens* and *Lygeum spartum*. It closes off 2 facies:

*Artemisia *campestris* and *Aristida pungens*

**Arthrocnemum indicum* and *Aerolopus littoralis*

2.4.4 Groups and facies of the Ouara

The plant groups and degradation facies of the Ouara are :

* Group with *Anthyllis sericea* ssp. *Heoniana* and *Gymnocarpos decander*. It is made up of 4 facies:

* *Anthyllis sericea* ssp. *Heoniana* and *Arthrophytum schmittianum*

* *Arthrophytum schmittianum* and *Helianthemum Lippii* var. *intricatum*

* *Arthrophytum schmittianum* and *Atractylis serratuloides*

* *Arthrophytum schmittianum* and *Anthyllis sericea* ssp. *heoniana*

-Sub-azonal group with *Retama raetam* and *Nitraria retusa*. It includes a single facies: facies with *Retama raetam* and *Polygonum equisetiforme*.

3. ECOLOGICAL SUMMARY

The climatic conditions of the decade of reallocation (1980-1990) classify the ecological regions of the Southern Low Plains, Matmata and J'fara under the **lower understorey arid Mediterranean bioclimate** type and the Ouara region under the **upper understorey Saharan Mediterranean bioclimate**. The thermal variants (winters) are, respectively, temperate, warm and mild and mild to warm. These types of bioclimate are those defined by Le Houérou (1959 and 1969) and Floret and Pontanier (1982). The rainfall regime of the Matmata and J'fara ecological regions, which open onto the Mediterranean, is of the **Autumn-Winter-Spring-Summer (AHPE)** type. It is of the **Winter-Autumn-Spring-Summer (HAPE)** type in the continental regions (Basses Plaines Méridionales and Ouara). So, whatever the geographical location of the regions, the rains fall in the cold season (Autumn or Winter) and the dry season, which begins just after the cold season, is a hot season. It lasts between 9 and 12 months a year.

The main characteristics of rainfall in these arid and desert areas are: **(i)** low amounts (95 mm/year to 200 mm/year), **(ii)** irregularity (rain-free years), **(iii)** high intensity (50 mm in one hour) and **(iiii)** poor distribution over the year (3 mm/month to 50 mm/month). Evapotranspiration is 1000 mm/year to 2700 mm/year and the dry period is very long (9 months/year to 12 months/year).

The dry climate (9 months/year to 12 months/year) in arid and desert areas is a constraint on the success of allochthonous species introduced for the reallocation of natural ecosystems. The low rainfall (3 mm/month to 50 mm/month maximum) and short duration (3 months/year to 4 months/year) do not match the monthly evaporative capacity of the soils (100 mm/month to 328 mm/month) and they do not coincide with the vegetative cycle of the introduced species (Cf. Chap. 1. § 1.4).

These climatic constraints are generally either accentuated or mitigated by the location of the ecosystem on the general toposequence (glacis) and by the intensity of the edaphic aridity (7 months/year to 9 months/year). This is a consequence of the geomorphological and physico-chemical characteristics of the soils and the low plant cover (10% to 30%).

Climatic aridity in glacis and slope ecosystems is generally accentuated by edaphic aridity due to the extent of run-off and poor infiltration of rainwater. On the other hand, in the ecosystems of plains, depressions and wadi beds, climatic aridity is relatively attenuated (rainwater and run-off).

The climatic and edaphic aridity that prevails in Tunisia's arid and desert zones are factors that limit the success of reallocation. Under these conditions of climatic and edaphic stress, the success of reallocation is most often dependent on rainfall conditions, the geomorphological situation and edaphic characteristics. The establishment, survival and production of introduced allochthonous species, and consequently the success of reallocation as a technique for creating new productive ecosystems, are more often than not problematic.

Chapter 3

RESPONSES of *Acacia saligna* TO ECOLOGICAL CONDITIONS OF REAFFECTED ECOSYSTEMS

Acacia saligna is the allochthonous shrub most widely used in the reallocation of natural ecosystems in arid and desert areas of Tunisia. It occupied between 45.3% and 96.6%, and on average 72.5%, of the areas reallocated in these zones during the 1980-1990 decade. The choice of *Acacia saligna* among ten or so introduced shrubs means that the areas occupied by the other shrubs (a few tens of square metres) do not allow for credible statistical analysis.

We note that the parameters for assessing the response of *Acacia saligna* to the ecological conditions of the reallocated ecosystems are :

-the success rate ;

-the production level of the average biovolume tree (epigated phytomass) ;

and that the statistical treatment is an analysis of variance followed by the Newman-Keuls test with a probability threshold of 5% (Cf. Chap. 1. § 2.3).

1. BEHAVIOUR IN RELATION TO GEOMORPHOLOGICAL AND EDAPHIC FEATURES

1.1. BEHAVIOUR IN SOUTHERN LOWLANDS ECOSYSTEMS

6 ecosystems, located in 4 ecological sectors, were reallocated in this region. Analysis of the variance in success rates and *Acacia saligna* production levels showed significant (alpha = 2%) and very highly significant (alpha = 0.1%) differences between ecosystems, respectively. Comparison using the Newman-Keuls test at the 5% probability threshold revealed two homogeneous groups for success rates and 4 groups for production levels (Tab. 12).

Table 12. Survival and production levels of *Acacia saligna* in the ecosystems of the Southern Lowlands. Comparison using the Newman-Keuls test (alpha = 5%)

<table>
<tr><th colspan="3">Plant success</th><th colspan="3">Production levels (Dehisced biomass of the average biovolume tree)</th></tr>
<tr><th>Type of ecosystem</th><th>Rate (%)</th><th>Group of homogeneous environments</th><th>Type of ecosystem</th><th>Kg of DM/year</th><th>Group of homogeneous environments</th></tr>
<tr><td>D1</td><td>69,0</td><td rowspan="4">A</td><td>P2</td><td>17,4</td><td>A</td></tr>
<tr><td>GC1</td><td>52,5</td><td>GC2</td><td>4,6</td><td rowspan="3">B</td></tr>
<tr><td>GG1</td><td>42,7</td><td>P1</td><td>2,8</td></tr>
<tr><td>GC2</td><td>31,7</td><td>D1</td><td>2,3</td></tr>
<tr><td>P1</td><td>18,0</td><td rowspan="2">B</td><td>GC1</td><td>1,1</td><td>C</td></tr>
<tr><td>P2</td><td>10,0</td><td>GG1</td><td>0,2</td><td>D</td></tr>
</table>

An analysis of success rates in the ecosystems of the Southern Lowlands region reveals two groups (Tab. 12). A first group with relatively high success rates (31.7% to 69%) and a second group with low success rates (10% to 18%).

The first homogeneous group is made up of ecosystems with little evolved soils with a very thick loose layer (60 cm to 100 cm) located either in ecological sectors of depressions (D1) or on limestone glacis covered by a thick to very thick loose layer (GC1). It should also be noted that ecosystems in gypsum glacis sectors with no soil cover (GG1) are also classified in this group. This can be explained by the safeguard work (replanting and watering) which lasts 3 years. The second group includes the ecosystems of the lowland sectors (P1 and P2) with a

very thick loose layer (>100 cm) covered by sharp aeolian morphologies (aeolian sails or dunes).
Success rates of over 50% are recorded in the ecosystems of depressions with soils of a thick loose layer with no identified substrate down to a depth of 100 cm (D1) and in limestone glacis with a thick loose layer (> 60 cm) with an impermeable substrate down to a depth of 100 cm (GC1). In these types of ecosystem, the geomorphological situation and the thickness of the loose layer create a favourable water regime: infiltration is equal to rainfall plus run-off.

Apart from ecological sectors with aeolian morphologies, the success rates of *Acacia saligna* in the ecosystems of the Southern Lowlands region would appear to be little affected by the geomorphological situation (glacis, depression, plain) and the physico-chemical characteristics of the soils (nature of the substrate, nature and thickness of the soft layer, limestone, salinity). The technique used to prepare the soil, plant the plants and, above all, protect and maintain the plants (replanting and watering) during the first three years accounts for this slight difference.

Sharp aeolian morphologies (aeolian sails and sharp dunes) bury young plants that do not have the characteristics of species adapted to these types of aeolian morphologies as defined by Bendali (1987).

Unlike success rates, *Acacia saligna* production levels in the reaffected ecosystems of the Southern Lowlands ecological region are influenced by geomorphological situation and edaphic characteristics. The location on the general toposequence (geomorphology), the thickness of the soft layer, the nature of the substrate and the wind morphology are the edaphic parameters that most determine the production of this shrub. Analysis of the production data reveals 4 groups of production levels (Tab. 11).

The first homogeneous group (P2), with the most remarkable production (17.4 kg DM/year), includes lowland ecosystems with very thick (>100 cm) loose layer soils covered by fixed aeolian morphologies (micro- or meso-nebkhas). The second group includes ecosystems with thick to very thick loose soil (60 cm to +100 cm) located in plains, depressions and on limestone glacis (2.3 kg DM/year to 4.6 kg DM/year) (GC2, P1 and D1). The third group comprises limestone glacis ecosystems (GC1) with soils of a thick to very thick loose layer (60 cm to 100 cm) and limestone glacis with no soil cover (GG1). Production is, successively, 1.1 kg DM/year 0.2 kg DM/year.

The presence of a fixed wind morphology (micro or meso-nebkas) on the ecosystems of the plains sectors whose soil is characterised by a thick to very thick loose layer (60 cm to 100 cm) improves the phytomass production of *Acacia saligna* by 4 to 17 times compared with the other types of ecosystem, whatever the nature of the substrate and the thickness of the loose layer. The capacity for infiltration and storage of water (rainwater and run-off) and the limitation of soil evaporation by the wind patterns explain this very high level of production (17.4 kg DM/year).

The factors limiting *Acacia saligna* production in the ecosystems of the Southern Lowlands region are :

- **the geomorphological situation: limestone or gypsum glacis with a thin layer (< 20 cm). In these types of ecosystems, rainwater run-off is significant and water infiltration is therefore limited;**
- **the outcrop of a limestone or gypsum crust on the surface;**

- **the amount of active limestone in the loose layer (> 12.6%);**
- **the gypsum content of the soil (> 23.2%);**

1.2. BEHAVIOUR IN MATMATA ECOSYSTEMS

Three ecosystems, located in a single sector (limestone glacis), were reallocated in the Matmata ecological region. Analysis of the variance in *Acacia saligna* success rates was insignificant at the 5% probability level (a single group). However, the difference in production levels was highly significant (alpha = 0.1%). The Newman-Keuls test of production levels at the 5% probability level reveals two homogeneous groups (Tab. 13).

Table 13. Survival and production levels of *Acacia saligna* in Matmata ecosystems. Comparison by Newman-Keuls test (alpha = 5%)

Plant success			Production levels (tree biomass of average biovolume)		
Type of ecosystem	Rate (%)	Group of homogeneous environments	Type of ecosystem	Kg of DM/year	Group of homogeneous environments
GC4	70,3		GC4	3,3	**A**
GC5	68,7	**A**	GC3	1,3	
GC3	60,0		GC5	1,0	**B**

The success rates of *Acacia saligna* in the ecosystems of this region vary from 60% to 70.3%, irrespective of the edaphic characteristics of the soils. The ecosystem where subsoiling and tabias are used (GC4) and the ecosystem where simple subsoiling is used (GC5) have the highest success rates, at 70.3% and 68.7% respectively. **Subsoiling and water and soil conservation work (tabias) have a remarkable influence on the survival rate of *Acacia saligna* (60% to 70%). The non-significant difference between the different ecosystems is due to the homogenisation of the water balance. The edaphic aridity in these limestone glacis ecosystems is thus mitigated by water and soil conservation work (subsoiling, creation of tabias). This work has encouraged the infiltration of rainwater and run-off and increased the soil's retention capacity.**

Acacia saligna production levels in the ecosystems of the Matmata region remain low (1 kg DM/year to 3.3 kg DM/year). However, they are influenced by the type of soil preparation work carried out. The highest production (3.3 kg DM/year) is recorded in ecosystems where water and soil conservation work (preparation of tabias) and subsoiling are practised (GC4). This production is around 3 times that of ecosystems treated by simple subsoiling (GC3) or without treatment (GC5). Simple subsoiling improves production compared with no treatment (1.3 kg DM/year compared with 1 kg DM/year). In addition to water and soil conservation work, the thickness of the loose layer (20 cm to 60 cm) and the presence of a substrate 60 cm deep also improve production levels.

Although subsoiling and water and soil conservation work have homogenised the soil's water balance and improved water infiltration, thereby reducing edaphic aridity, *Acacia saligna* production remains very low (< 3 kg DM/year). This low production can be explained by the active limestone content of the soil (> 15%). This shrub is therefore sensitive to limestone. *Acacia saligna* is therefore a calcareous species.

1.3. BEHAVIOUR IN J'FARA ECOSYSTEMS7

Reallocation in this region involved 12 ecosystems located in 5 ecological sectors. Analysis of variance of the data shows a highly significant difference (alpha = 0.4%) between the success rates of *A. saligna* and a very highly significant difference (alpha = 0.1%) between the production levels.

Analysis of the success rates in the ecosystems of the J'fara region distinguishes 3 groups (Tab. 14). The first homogeneous group is characterised by very high success rates (81% to 98.5%), the second by relatively high success rates (36.1% to 63.3%) and the third by low success rates (19.3% to 25.5%).

Analysis of production levels, on the other hand, reveals 6 groups (Tab. 14). The first group and the second are each made up of a single type of ecosystem, with production of 10 kg DM/year and 6 kg DM/year respectively. The third homogeneous group is made up of three types of ecosystem with production ranging from 3.7 kg DM/year to 4.7 kg DM/year. The fourth group comprises a single ecosystem (2.7 kg DM/year). Finally, the fifth and sixth groups are made up of three types of ecosystem, with production ranging successively from 1.3 kg DM/year to 1.5 kg DM/year and from 0.1 kg DM/year to 0.3 kg DM/year.

Table 14. Survival and production levels of *Acacia saligna* in the J'fara ecosystems. Comparison by Newman-Keuls test (alpha = 5%)

Plant success			Production levels: (Dehisced biomass of the average biovolume tree)		
Type of ecosystem	Rate (%)	Group of homogeneous environments	Type of ecosystem	kg DM/year	Group of homogeneous environments
GC8	95,8	A	GC8	10,0	A
P5	90,0		D3	6,0	B
BMT2	87,5		P4	4,7	C
P4	81,0		BMT1	4,1	
P3	63,3	B	GC7	3,7	
D2	50,3		P3	2,7	D
GC6	45,7		D2	1,5	E
BMTl	42,2		BMT2	1,3	
GC7	36,1		GC6	1,3	
GG2	25,5	C	P5	0,3	F
D3	20,0		GG2	0,2	
D4	19,3		D4	0,1	

The majority of ecosystems in the J'fara region are characterised by a success rate of over 40%. These high rates, as already mentioned for the ecosystems in the previous ecological regions, are the result of the soil preparation and plant maintenance and protection work (watering and replanting) carried out during the first three years after planting. This work helped to replace missing plants and to homogenise soil water reserves. In halomorphic ecosystems or in soils with a loose saline layer, periodic watering over a three-year period has enabled salts to be leached out at depth.

However, the level of production of *Acacia saligna* in the ecosystems of this region is linked to the physico-chemical characteristics of the soil: thickness of the loose layer, presence of an impermeable substrate at a certain depth, presence or absence of limestone, gypsum or salts. The most favourable ecosystems for the production of this allochthonous shrub are those characterised by a thick to very thick (60 cm and 100 cm) loose layer of soil on an impermeable substrate, preferably limestone. Ecosystems with soils without a substrate, even if the loose layer is very thick (> 100 cm) and those characterised by the presence of a limestone or gypsum substrate at a depth of less than 60 cm, are not very favourable for *Acacia saligna* production.

The factors limiting *Acacia saligna* production in the J'fra region are :

- **the presence of a limestone or gypsum substratum at a depth of less than 60 cm (screen for water percolation at depth);**
- **the absence of an impermeable substrate at a depth of 100 cm (percolation of water at depth);**
- **soils rich in active limestone (> 15%);**
- **the salinity of the soft layer (> 5 mS/cm) whatever its thickness;**
- **halomorphic soils.**

***Acacia saligna* is therefore a shrub that does not tolerate halomorphic soils. It fears salts and active limestone in the loose layer, however thick it may be.**

1.4. BEHAVIOUR IN OUARA ECOSYSTEMS

Five ecosystem types, located in 3 sectors, were reallocated within this ecological region. Analysis of the variance in survival rates and production levels of *Acacia saligna* shows very highly significant differences (alpha = 0.1%) between ecosystems (Tab. 15). The groups of homogeneous environments discriminated by the Newman-Keuls test at the 5% probability threshold are 3 for each of these parameters (Tab. 15).

Table 15. Survival and production levels of *Acacia saligna* in Ouara ecosystems. Comparison by Newman-Keuls test (alpha = 5%)

<table>
<tr><th colspan="3">Plant survival</th><th colspan="3">Production levels: (Tree biomass of average biovolume)</th></tr>
<tr><th>Type of ecosystem</th><th>Rate (%)</th><th>Group of homogeneous environments</th><th>Type of ecosystem</th><th>Kg of DM/year</th><th>Group of homogeneous environments</th></tr>
<tr><td>GG4</td><td>64,7</td><td rowspan="2">A</td><td>BMT3</td><td>3,4</td><td>A</td></tr>
<tr><td>GG3</td><td>54,3</td><td>GG3</td><td>1,9</td><td>B</td></tr>
<tr><td>BMT3</td><td>29,0</td><td rowspan="2">B</td><td>GG4</td><td>0,6</td><td rowspan="3">C</td></tr>
<tr><td>GC10</td><td>23,3</td><td>GC10</td><td>0,2</td></tr>
<tr><td>GC9</td><td>16,0</td><td>C</td><td>GC9</td><td>0,1</td></tr>
</table>

Acacia saligna survival rates are high (64.7% and 54.3%) in gypsum crust ecosystems (GG4 and GG3). In the ecosystems of the low and medium terraces (BMT3) and the limestone glacis with very thick (GC10) and thick (GC9) loose layers, survival rates are low to very low, successively 29%, 23.3% and 16%.

The low survival rates of *Acacia saligna* in the ecosystems of low and medium terraces and limestone glacis covered by a thick to very thick loose layer do not reflect the reality of the ecological response of this shrub to the favourable ecological conditions of these ecosystems, as is the case in other ecological regions. In fact, the sustained anthropozoic actions (illegal cutting and grazing) exerted on these ecosystems are the causes of these low success rates.

Unlike survival rates, geomorphological and edaphic characteristics have an effect on *Acacia saligna* production in Ouara ecosystems. However, it remains very low, ranging from 0.1 kg DM/year to 3.4 kg DM/year, whatever the geomorphological and edaphic characteristics of the ecosystem (Tab. 15). The most favourable ecosystems for the production of this shrub are the low and medium terraces of wadis with a loose layer more than 100 cm thick (3.4 kg DM/year) and the gypsum glacis covered by a loose layer of soil 20 cm thick (1.9 kg DM/year). **Does this mean that *Acacia saligna* exerts an extraction force greater than the water retention force of the gypsum crust?**

Production in ecosystems on limestone glacis, whatever the thickness of the loose layer, remains very low (< 0.6 kg DM/year).

Although the geomorphological and edaphic conditions of the Ouara ecological region are similar to those of the other three regions, the level of production of *Acacia saligna* remains very low (< 3.4 kg DM/year). It would therefore appear that climatic conditions (low rainfall, high temperatures) and high evaporation and evapotranspiration are the factors limiting the production of this shrub.

2. BEHAVIOUR IN RELATION TO BIOCLIMATIC CHARACTERISTICS

The reaffected ecosystems of Tunisia's arid and desert zones are influenced by two types of Mediterranean bioclimate: lower arid and upper Saharan.

2.1. BEHAVIOUR IN THE LOWER ARID BIOCLIMATE

The reallocated ecosystems subject to this type of bioclimate with a temperate to hot variation are those of the ecological regions: the Southern Low Plains, the Matmata and the J'fra. The salient features of the climatic conditions during the decade 1980-1990 (see Tab. 6 and 7) are :

- poor intra- and inter-annual distribution;
- a succession of dry years: 2 years out of 3 ;
- severe and long-lasting climatic and edaphic drought, successively 9 months/year to 11 months/year and 7 months/year to 9 months/year;
- low monthly rainfall (0.7 mm to 41.4 mm);
- autumn or winter rainfall;
- a high temperature and a high thermal amplitude;
- very high monthly evaporative demand (87 mm to 375.6 mm);

These climatic characteristics, which are relentless even for native species, mean that *Acacia saligna* is subject to permanent climatic stress. This type of stress is accentuated by edaphic and physiological stress due to the physico-chemical characteristics of the soil.

In the arid Mediterranean understorey bioclimate, the survival of *Acacia saligna* after 3 years of maintenance work (watering and replanting) generally depends on the geomorphological situation of the ecosystem. Plains, depressions, low and medium terraces and glacis covered by a thick loose layer (> 60 cm) are the most favourable for the establishment and success of this shrub. The survival rate of *Acacia saligna* in these conditions is over 50% (Tab. 16). The location on the general toposequence (plain) and the physical characteristics of the soil (thickness of the soft layer) favour infiltration and storage of water and improve the water balance (rainwater and run-off).

The presence of ecosystems with different geomorphological and edaphic characteristics in the same statistically homogeneous group (plains, depressions, limestone glacis, gypsum glacis) or of the same type of ecosystem in statistically different groups (Tab. 16) indicates that the establishment and survival of *Acacia saligna* are very little influenced by geomorphological and edaphic factors. **Can we therefore accept that *Acacia saligna*, in the lower arid Mediterranean bioclimate, is characterised by a wide edaphic amplitude and plasticity with respect to these factors during the establishment phase?**

The wide range and plasticity of *Acacia saligna* in relation to geomorphological and edaphic factors, in this type of bioclimate, are in fact the result, on the one hand, of the soil preparation work and, on the other hand, of the maintenance and protection of the seedlings (replanting and watering) which are carried out from the time the seedlings are planted until they are 3 years old. Soil preparation removes the crust and the calcareous or gypseous encrustation, replanting replaces missing plants and periodic watering makes up for the lack of water.

Nor does soil salinity have a negative influence on the survival rate of *Acacia saligna* in the lower arid Mediterranean bioclimate. In fact, in halomorphic formations (conductivity of the order of 20 mS/cm in the hot dry season), the survival rate of this shrub is very high (> 80%). Neffeti et *al* (1990) reported the same phenomenon of resistance of this shrub to salinity during the establishment phase. **Can we therefore consider *Acacia saligna* to be a salinity-tolerant shrub?**

Table 16. Survival rate and production level *of Acacia saligna* in the lower arid Mediterranean bioclimate

Plant survival						Production levels: (Dehisced biomass of the average biovolume tree)					
Ecosystem	Ecological region	Average rainfall mm/year	Average evap. mm/year	Success rate % (%)	Group of homogeneous environments	Ecosystem	Ecological region	Average rainfall mm/year	Average evapo. mm/year	Producer. kg DM/year	Group of homogeneous environments
GG2				95,8		P2	Southern Lowlands	173,6	2471,2	17,4	A
P5	J'fara	183,1	1 979,2	90,0	A	GC8				10,0	B
BMT2				87,5		D3	J'fara	183,1	1 979,2	6,0	
P4				81,0		P4				4,7	C
GC4	Matmata	195,0	2 040,7	70,3		GC2	Southern Lowlands	173,6	2471,2	4,6	
Dl	J'fara	183,1	1 979,2	69,0		BMTl				4,1	
GC5	Matmata	195,0	2 040,7	68,7	B	GC7	J'fara	183,1	1 979,2	3,7	
P3	J'fara	183,1	1 979,2	63,3		GC4	Matmata	195,0	2 040,7	3,3	
GC3	Matmata	195,0	2 040,7	60,0		Pl	Southern Lowlands	173,6	2471,2	2,8	D
GCl	Southern Lowlands	173,6	2471,2	52,5		P3	J'fara			2,7	
D2	J'fara	183,1	1 979,2	50,3	C	Dl				2,3	
GC6				45,7		D2		183,1	1 979,2	1,5	E
GGl	Southern	173,6	2471,2	42,7		GC3	Matmata	195,0	2 040,7	1,3	

BMTI	Lowlands			**42,2**		**BMT2**	J fara	183,1	1 979,2	**1,3**	
GC7	J'fara	183,1	1 979,2	**36,1**		**GC6**				**1,3**	
GC2	Southern Lowlands	173,6	2471,2	**31,7**	**D**	**GCI**	Southern Lowlands	173,6	2471,2	**1,1**	
GC8	J fara	183,1	1 979,2	**25,5**		**GC5**	Matmata	195,0	2 040,7	**1,0**	
D3				**20,0**	**E**	**P5**	J'fara	183,1	1 979,2	**0,3**	**F**
D4				**19,3**		**GGI**	Southern Lowlands	173,6	2471,2	**0,2**	

This hypothesis is invalidated by the fact that periodic watering during the first three years allows salts to be leached out at depth. Stopping watering after the third year of planting leads to dieback of *Acacia saligna* installed in these types of saline soil (EC > 5 mS/cm) and halomorphic soils (personal observations).

Finally, we note that the presence of mobile aeolian formations (aeolian sails, sand sheets, micro- and meso-dunes) has a negative impact on the success rate (10% to 25%). These formations bury the young plants, which do not have the characteristics of species adapted to sharp wind morphologies.

The level of production of *Acacia saligna* in the lower arid Mediterranean bioclimate type, unlike survival, depends on the geomorphological situation and the physico-chemical characteristics of the soils (Tab. 16). Plains ecosystems with a very thick loose layer (> 100 cm) and calcareous glacis ecosystems with a thick loose layer (60 cm to 100 cm) covered by fixed aeolian morphologies (micro- and meso-nebkas) provide the highest level of production: 17.4 kg DM/year and 10 kg DM/year (Tab. 16). Fixed aeolian morphologies quadrupled the production level of *Acacia saligna* compared with an ecosystem with a loose layer of the same thickness but without aeolian morphology: 17.4 kg DM/year compared with 4.6 kg DM/year (Tab. 16).

In limestone glacis ecosystems, even if the loose layer is thick (60 cm to 100 cm), and in gypsum glacis ecosystems with no loose layer, *Acacia saligna* production is compromised by high levels of active limestone or gypsum in the soil. Production levels in these types of ecosystem are 1.1 kg DM/year and 0.2 kg DM/year respectively. Unlike survival, soil salinity is the limiting factor in *Acacia saligna* production. The production levels of this shrub in saline ecosystems and halomorphic soils in the lower arid Mediterranean bioclimate are statistically no different from those recorded in ecosystems with no or a thin loose layer (0 cm to 60 cm).

While the position of the ecosystem on the general toposequence (geomorphological situation) and its physico-chemical characteristics under the lower arid Mediterranean bioclimate do not have a very significant influence on the success rate and consequently the establishment of *Acacia saligna* due to the work involved in preparing the soil and maintaining and safeguarding the seedlings ; However, they do accentuate or attenuate water stress and, consequently, have an effect on the production level of this allochthonous shrub, which has been introduced for the reallocation of ecosystems in the arid Mediterranean bioclimate in general.

2.2. BEHAVIOUR IN THE UPPER SAHARAN BIOCLIMATE

The upper understorey Saharan Mediterranean bioclimate covers the Ouara region. The

ecological characteristics during the decade 1980-1990 (see Tab. 6 and 7) were as follows:
- low average annual rainfall (95.2 mm) and monthly rainfall ranging from 1 mm to 21.1 mm;
- winter rainfall;
- very high average annual evaporation (2721.8 mm) and monthly evaporation ranging from 121.1 mm to 329.9 mm ;
- a dry period of 12 months/year.

These climatic conditions, accentuated by edaphic aridity, are too far removed from the requirements of *Acacia saligna*, which has been introduced to rehabilitate ecosystems in this type of bioclimate.

The survival and establishment of *Acacia saligna* in the upper Saharan Mediterranean bioclimate type is conditioned by the edaphic characteristics of the ecosystem (see Tab. 15), unlike in the lower arid Mediterranean bioclimate. Ecosystems with gypsum substrates are the most favourable for the establishment of *Acacia saligna* (54.3% to 64.7%). On the other hand, ecosystems with a limestone substrate, even when subsoiled, do not seem to be very conducive to the establishment and survival of this shrub (16% to 23%). The ecosystems of the low and medium terraces of the wadis, in the absence of human activity, are the most favourable to the establishment and survival of this shrub.

Although *Acacia saligna* production in the upper Saharan Mediterranean bioclimate is very low (maximum 3.3 kg DM/year), it depends on the edaphic conditions of the ecosystem (see Tab. 15). The ecosystems that are potentially favourable for the production of this shrub are the low and medium terraces of the wadis. Ecosystems with a very thick loose layer (> 100 cm), which could also be considered as potentially productive, only allow, however, a very low production (0.2 kg DM/year). This production is not significantly different from that recorded in ecosystems without a loose layer or with a thin layer (0 cm to 60 cm).

Despite the favourable edaphic conditions, plant production of *Acacia saligna* in the Saharan bioclimate is very limited (see Tab. 15). This is due to the severe climatic constraints of this type of bioclimate: low rainfall, high thermal amplitude, very high evaporation and a dry period throughout the year.

In the Saharan Mediterranean bioclimate, climatic conditions rather than geomorphological situation and edaphic characteristics are the limiting factors for the establishment, success and production of *Acacia saligna*.

GENERAL DISCUSSION AND EVALUATION OF THE REALLOCATION

Tunisia's arid and desert zones are exposed to three types of stress:

- **climatic stress** due to **(i)** low rainfall, **(ii)** very high evaporative demand and **(iii)** climatic drought from 9 months/year to 12 months/year;
- **edaphic stress** (7 months/year to 9 months/year) which is a consequence of the geomorphological situation and the physical characteristics of the soils: glacis, crust and superficial limestone or gypsum or calcairo-gypsum crusts.
- **physiological stress** due to the chemical characteristics of the soil: salts, gypsum and limestone. In addition to these permanent stresses (climatic, edaphic and physiological), the majority of introduced allochthonous shrubs, including *Acacia saligna*, have a spring and summer development cycle. This cycle is out of sync with the rainfall patterns of Tunisia's arid and desert zones (Autumn-Winter-Autumn-Spring or Winter-Autumn-Spring-Summer).

Thus, the ecological conditions of Tunisian arid and desert zones are far from meeting the requirements for the establishment, growth and production of allochthonous shrubs.

The ecosystems earmarked for reallocation in these areas, where climatic aridity and physiological stress are omnipresent throughout the year, are generally characterised by erosion glacis (limestone or gypsum), poor soils with a superficial indurated horizon and soils that are not conducive even to the success and production of native species due to edaphic aridity. In fact, more than 2/3 (70%) of the surface area of ecosystems earmarked for conversion in arid and desert areas of Tunisia are located on limestone or gypsum glacis. Geomorphological situations that are potentially favourable for the establishment, survival and production of introduced shrubs (plains, deep soils, depressions and wadi terraces) account for only around ¼ (24%).

While the geomorphological situation of the ecosystem is a major constraint to successful reallocation, the thickness of the loose layer can accentuate or attenuate this constraint and consequently accentuate or attenuate edaphic drought. In arid and desert areas of Tunisia, approximately 2/3 (61.9%) of the soils in reallocated ecosystems are characterised by a thin loose layer (< 60 cm). As a result, they are not conducive to the establishment, success and production of introduced shrubs.

So what will happen to the ecosystems in Tunisia's arid and desert zones?

In the arid Mediterranean and Saharan bioclimate, the establishment and survival of *Acacia saligna* are relatively independent of the nature of the soil in the ecosystem. In other words, the edaphic characteristics (nature and thickness of the loose layer, nature of the substrate, salinity) do not have a major influence on the establishment and survival of this shrub. The same is true for all introduced species.

Could this behaviour be indicative of the "plasticity" of these allochthonous shrubs in Tunisia's arid desert zones?

This assumption should not conceal the impact of soil preparation work and, above all, maintenance and protection work (watering and replanting) during 3 consecutive years of planting. This work masks the influence of the geomorphological situation and the physico-chemical characteristics of the soil on the establishment and survival of allochthonous shrubs introduced under arid Mediterranean and Saharan bioclimates, and even under semi-arid bioclimates. In these types of bioclimate, the establishment and survival of allochthonous shrubs, and consequently the success of ecosystem reallocation, are compromised by the geomorphological situation and the following edaphic characteristics:

-Ice
-more or less superficial limestone or gypsum **substrate** (loose layer < 20 cm thick) ;
mobile aeolian morphologies (sand blanket, aeolian veil, micro- and meso-dunes) which bury young plants.

Halomorphic soils and soils with loose saline layers appear to be favourable environments for *Acacia saligna*. **Does this mean that this shrub tolerates salts?** Neffeti et *al* (1990) and Zaâfouri et *al* (1991 and 1994), Zaâfouri (1993) have reported that *Acacia saligna* can withstand salt levels of around 20 mS/cm in the summer during the establishment phase. **This tolerance, not resistance, is in fact a consequence of periodic and prolonged watering during the first three years after planting**. This water leaches salts from the surface horizons (0 cm to 80 cm), carrying them deep down. If the watering stops, the standing *Acacia saligna* and any introduced species that cannot tolerate salts will wither and die.

The production of allochthonous shrubs introduced to the arid Mediterranean bioclimate under the lower storey, unlike their success, is linked to the geomorphological situation and the physico-chemical characteristics of the ecosystem soils. The highest production of the average biovolume *Acacia saligna* tree (17.4 kg DM/year) is observed in the ecosystems of the plains sectors where the thickness of the loose layer is 150 cm covered by meso-nebkas. This production is much higher than that recorded under the arid Mediterranean bioclimate in the upper understorey on sediments (2.6 kg DM/year) (Akrimi et *al.*, 1989) or on a very thick (> 100 cm) loose layer of silty-sandy texture (10.8 kg DM/year) (Ktata, 1986). In the Saharan Mediterranean bioclimate, the production of this shrub is very low, whatever the ecological conditions of the ecosystem (0.1 kg DM/year to 3.4 kg DM/year).

So the question is: is it possible to reallocate the natural ecosystems of Tunisia's arid and desert zones using non-native shrubs?

In the arid Mediterranean bioclimate, it is generally possible to reallocate ecosystems by, but in geomorphological situations and with particular edaphic characteristics, which are :

- position on the general toposequence: plains, depressions, low and medium terraces and wadi beds;
- soil cover: a loose layer at least 60 cm thick for ecosystems located on glacis;
- the presence of an impermeable substrate at a depth of between 60 cm and 100 cm ;
- a limestone or gypsum substratum surmounted by fixed wind patterns ;
- low levels of active limestone and gypsum;
- a neutral to slightly basic pH;
- very low salinity of the soft layer (EC < 5mS/cm).

Unfortunately, land with these characteristics is most often reserved for subsistence farming (market gardening, arboriculture and cereals). There are, however, very few options left for the successful reallocation of ecosystems through the introduction of allochthonous shrubs.

In the Saharan bioclimate, the reallocation of natural ecosystems is more problematic: a relatively high success rate due to maintenance and conservation work during the first few years, but very low production.

We conclude that the bioclimatic limit for the successful reallocation of ecosystems in arid and desert zones in Tunisia is the lower understorey of the arid Mediterranean bioclimate. The rainfall is at least 150 mm/year. These bioclimatic and climatic conditions must prevail over geomorphological situations and edaphic characteristics that are favourable to plant production. In addition, these conditions are increasingly being used to grow food crops

such as arboriculture and irrigated market gardening.

The success of the technique for reallocating degraded or desertified natural ecosystems in arid and desert areas is generally problematic: the investment cost for soil preparation, planting and maintenance and conservation work is high. On the other hand, the installation, success and production of allochthonous shrubs introduced into these arid and desert conditions in Tunisia remain very limited.

BIBLIOGRAPHY

Akrimi N, 1986 -Rapport de mission au Brésil sur le *Prosopis* (25-28/9/1986). International Conference on Prosopis. Document interne de l'Institut des Régions Arides, Médenine, Tunisie : 97 p.

Akrimi N et **Zaâfouri M.S**, 1990 -Etude des arbustes fourragers les plus couramment utilisés dans la mise en valeur des régions arides tunisiennes : Description, écologie et techniques d'exploitation. Revue des Régions Arides 1/90. Editions Institut des Régions Arides Médenine, Tunisia: pp. 4-83.

Akrimi N, **Zaâfouri M.S** and Neffeti M, 1995 -Végétation steppique nord-africaine. Séminaire International. Relation steppe-oasis dans les zones arides nord saharienne. (MED-CAMPUS). Université P. et M. Curie Paris/Université de Caen. France/Institut Régions Arides Médenine, Tunisia/N.S.R.C. Giza Cairo : 10 p.

Akrimi N, **Zaâfouri M.S**, Romdhanne A et Jeder H, 1996 -Caracterisation des ressources naturelles et des populations de la zone du Parc de Haddej. Editions de l'Observatoire du Sahara et du Sahel (OSS): 40 p.

Amrani S, Noureddine N-E and Bhatnagar T, 2009 -Symbiotic and genotypic characteristics of Rhizobia associated with *Acacia saligna* (Labill.) Wendl. in some nurseries in Algeria. Acta Bot. Gallica. 156 (3) : pp. 501-513.

APG II (Angiosperm Phylogeny Group II), 2003 -Angiosperm Phylogeny Group -An update of the Angiosperm Phylogeny Group classification for the orders and families of flowering plants. Botanical Journal of the Linnaean Society. Vol. 141 (4) : pp. 399-436.

APG III (Angiosperm Phylogeny Group III), 2009 -Angiosperm Phylogeny Group -An update of the Angiosperm Phylogeny Group classification for the orders and families of flowering plants. Botanical Journal of the Linnean Society. Volume 161 (2) : pp. 105-121.

Aronson J, Floret Ch., Le Floc'h E, Ovalle C and Pontanier R, 1995 -Restoration and rehabilitation of degraded ecosystems in arid and semi-arid zones. In l'Homme peut-il refaire ce qu'il a défait. Jean Libbey Editions, Paris, France: pp. 11-29.

Asocar R, Mansilla A and Silva H, 1981: Método de estimacion de la fitomassa ytili de *Atriplex repanda*. Av. Prod. Anim. vol. 6 (1-2): pp 21-28.

Attia A, 1977 -Les hautes steppes tunisiennes : de la société pastorale à la société paysanne. Doctorat ès Lettres thesis. University of Paris VII, France. 400 p.

Auclair L et **Zaâfouri M.S**, 1996 -La sédentarisation des nomades dans le sud tunisien : comportements énergétiques et désertification. Sécheresse. Volume 7. N° 1. John Libby Editions, Paris, France: pp. 1-8.

Barbéro M, 1981 -Les frutacées de la région de la zone bioclimatique méditerranéenne à *Chêne pubscent* : structure, zonage, utilisation, protection et biomasse. Forêts Méditerranéennes. III (2): pp 101-114.

Baudin F, 1985 -Phytovolume, phytomass and spatial occupation strategy of six shrub species on Esterel firebreaks. D.E.A d'Ecologie. Université d'Aix-Marseille III, France: 34 p.

Bendali F, 1987 -Dynamique de la végétation et mobilité du sable en Jeffara tunisienne. Doctoral thesis. Faculté des Sciences et Techniques Languedoc, Montpellier, France: 243 p.

Bonneau M et Timbal J, 1973 -Définition et cartographie des stations. French and foreign concepts. Annales des Sciences Forestières. Volume 30 (3) : pp 201-218.

Boudy P, 1950 -Economie forestière nord-africaine. Monograph and treatment of forest

species. Tome 2. Fascicule 1. Editions La Rose: 525 p.
Bourges J, Floret C et Pontanier R, 1975 -Etude d'une toposéquence type du sol tunisien "Djebel Dissa" : Les sols, ruissellement, bilan hydrique, érosion et végétation. Direction des Ressources en Eaux et en sols, Minis. Agriculture, Tunisia. ES N° 93; 56 p.
EEC (European Economic Community), 1997 - Rehabilitation of degraded land north and south of the Sahara. Use of perennial legumes and associated micro-organisms for the establishment of pluristrate formations. Final scientific report. Centre d'Etudes Phytosociologiques et Ecologiques Louis Emberger (CEPE), Montpellier, France. Montpellier, France/Office de Recherche Scientifique et Technique Outre-mer, Mission de Tunisie/Institut des Régions Arides, Médenine, Tunisia. 164 p.
Chaieb M, 1989 -Influence of soil water reserves on the comparative behaviour of some plant species in the Tunisian arid zone. Doctoral thesis in Science. Université des Sciences et Techniques Languedocienne, Montpellier, France: 293 p.
Chaieb M, 1991 -Steppes tunisiennes : Etat actuel et possibilités d'amélioration. Sécheresse N° 2. Volume 2: pp. 95-99.
Cuénot A, in collaboration with Pottier-Alapetite G and Labbé A, 1954 -Flore analytique et synoptique de la Tunisie. Vascular cryptogams, gymnosperms and monocotyledons. Imprimerie S.E.F.A.N : 287 p.
Chouaieb T, 1982 -Détermination de la production fourragère d'*Acacia cyanophylla* dans les périmètres sylvo-pastoraux de la zone aride tunisienne. Document interne de la Direction des Forêts, Ministère de l'Agriculture, Tunisie : 7 p.
Clément B et Touffet J, 1976 -Biomasse végétale aérienne et productivité des moes des Monts d'Arrée (Bretagne, France). Acta Oecologica. Oecologya Plantarum. Volume 11 (4) : pp 345360.
Coque R, 1962 -La Tunisie présaharienne : Etude géographique. Armand Colin Editions, Paris, France: 488 p.
Daget Ph, 1977a -Le bioclimat méditerranéen : Caractéristiques générales et modes de caractérisation. Végétation. Volume 34 (1): 1-20.
Dagnelie P, 1975 - Statistical theory and methods. Agronomic applications. Volume II: Les méthodes de l'inférence statistique. Edit. Presse Agron, Gembloux. A.S.B.L., Belgium : 405 p.
Decourt N, Godron M, Romaine F et Tomassonne R, 1969 -Comparaison de diverses méthodes d'interprétation statistique de liaison entre le milieu et la production su Pin sylvestre en Sologne. Annales des Sciences Forestières, France. Volume 6 (4): pp. 413-443.
Decourt N, 1973 a -Production primaire, Production utile : Méthode d'évaluation, indice de productivité. Annales des Sciences Forestières, France. Volume 30 (3): pp 219-239.
Dumancic D and Le Houérou H.N, 1981, *Acacia cyanophylla* as supplementary feed for sall stock in Libya. Journal of arid environment. Volume 4 : pp. 161-167.
Duvigneaud P, 1974 -La synthèse écologique : Populations, communautés, écosystèmes, biosphère et nanosphère. Editions Doin Paris, France: 296 p.
El Hamrouni A, 1992 -Végétation forestière et préforestière de la Tunisie. Typologie et éléments pour la gestion. Revue des Régions Arides, Médenine, Tunisia. N° spécial 6/94. Editions Institut des Régions Arides, Médenine, Tunisia: 299 p.
El Hamrouni A et Sarson M, 1975 a -Développement des cultures d'*Atriplex* en Tunisie depuis 1971. Bulletin d'Information N° 18. Institut National de Recherche Forestière, Tunisia.

: 20 p.
Emberger L, 1955 -Une classification biogéographique des climats. Revue et Travaux de Laboratoire de Botanique et de Zoologie. Service de Botanique. N° 7. Faculté des Sciences, Montpellier, France: pp. 3-43.
Etienne M, 1980 -Non destructive methods for evaluating shrub biomass. Acta Oecologica. Oecologica Plantarum. Volume 10 (2) : pp 115-128.
Evans F.C, 1956 -Ecosystems as the basic unit. Ecology Sciences. 123 : pp. 1127-1127.
FAO, 1982 - Experimental facilities for species and provenance testing: Establishment and evaluation. Basic Instruction Document II. FAO Project on Genetic Resources of Arid and Semi-arid Zones. FORM/9/1980: 220 p.
Felker P, Cannel G, Osborn J and Clarrk P, 1982 -Biomass estimation a young stand of Mesquite (*Prosopis*), ironwood (*Oluega testa*) palo verde an *Leucaena*. Journal of Range Management. Volume 35 (1) : pp 87-89.
Ferchichi A, 1999 -Les parcours de la Tunisie présaharienne : Potentialités, état de désertification et problématique d'aménagement. Options Méditerranéenne. IAM Zagros/Centre des Hautes Etudes Agronomiques Méditerranéennes. No. 39: pp. 137-141.
Floret Ch., Le Floc'h E, Romane, F, Le Part J and David P, 1973 -Production, sensitivity and evolution of natural vegetation and the environment in pre-Saharan Tunisia. Conséquence pour la planification. Institut National de Recherche Agronomique, Tunis, Tunisia/Centre d'Etudes phytosociologiques et Ecologiques Louis Emberger, Montpellier, France. Document N° 71 : 45 p.
Floret Ch., Le Floc'h E and Pontanier R, 1983 - Phytomass and plant production in pre-Saharan Tunisia. Acta Ecologia. Ecologia Plantarum. Volume 4 (18). No. 2: pp. 133-152.
Floret Ch., 1988 -Methods for measuring pastoral vegetation. Pastoralism and development. Institut Agronomique et Vétérinaire Hassen II, Rabat, Morocco/Institut Agronomique Méditerranéen, Montpellier, France : 157 p.
Floret Ch. et Pontanier R, 1982 -L'aridité en Tunisie présaharienne. Climat, sol, végétation et aménagement. Travaux de l'ORSTO). Editions Office de Recherche Scientifique et Technique Outre-mer (ORSTOM), Paris, France: 543 pp.
Floret Ch. et Pontanier R, 1984 -Aridité climatique, Aridité édaphique. Bulletin des Sciences Botaniques Française. N° 2/3/4 : pp. 265-275.
Feuillas D, 1979 - Méthodes et techniques d'estimation de la biomasse épigée des formations arbustives et leurs applications au maquis corse dans la vallée du Fargo. D.E.A d'Ecologie Végétale. University of Paris-Sud, France: 51 p.
Franclet A et Le Houérou H.N, 1971 -Les Atriplexes en Tunisie et en Afrique du Nord. Rapport Technique. N° 7. FAO, Rome, Italy. 178 p.
Gammar A.M, 1989 -Paléophlore tunisiennes et mise en place de la végétation actuelle. In. Essai de synthèse sur la végétation et la phytoécologie tunisiennes. I : Eléments de botanique et phytoécologique. MAB Programme: Tunisian Flora and Vegetation. Volumes 4 to 6. Faculté des Sciences de Tunis, Tunisia: pp. 195-233.
Garbaye J, Le Roy P, Le Facon F et Levy G, 1970 -Réflexion sur une méthode d'étude des relations entre facteurs écologiques et caractéristiques des peuplements. Annales des Sciences Forestières, France. No. 27: pp. 304-317.
Godron M et Poissonet J, 1972 -Four complementary themes for mapping vegetation and the

environment: vegetation sequence, landscape diversity, healing speed, vegetation sensitivity. Bulletin de la Société Languedocienne de Géographie, Montpellier, France. Volume 6 (3): pp. 329-359.

Gounot M, 1969 -Méthodes d'études quantitatives de la végétation. Masson et Cie Editions, Paris, France: 314 p.

Hamza H, 1986 -Research on *Acacia* in Tunisia to improve forestry and pastoral production and combat desertification. Séminaire International sur la lutte contre la désertification. Jerba, Tunisia 24-28/11/1986 : 7 p.

Jeder H, **Zaâfouri M.S**, Boukhchina R, Romdhane A, Talbi M, Ouled Belgacem A et Znati S, 2000 -Suivi à long terme de la biodiversité et de la population dans l'observatoire pilote de Haddej- Bou Hedma, Tunisie. Scientific Report. 1ère phase du Projet Réseau de Suivi d'Observatoire de Surveillance Ecologique à Long Terme (ROSELT). Institut des Régions Arides, Médenine, Tunisia/ROSELT/Observatoire du Sahara et du Sahel (OSS): 50 p.

Karem A, 1984 -Contribution à l'étude du bilan fourrager des interventions dans les périmètres de Ouled Chraiet (Gafsa) : Evaluation de la production des parcours et des plantations de *Cactus inerme* et leur contribution à l'alimentation des animaux. Dissertation 3ème cycle. Institut Agronomique et Vétérinaire Hassen II, Rabat, Morocco: 137 p.

Karem A et Ktata T, 1986 -Etude de biomasse : détermination de la production fourragère par hectare de certaines espèces pastorales dans divers périmètres de parcours améliorés des Gouvernorats de Sidi Bouzid, Sfax, Kasserine et Mahdia. Rapport Technique. Direction des Forêts, Minis. Agriculture, Tunisia: 14 p.

Lamotte M, 1962 -Initiation aux méthodes statistiques en biologie. 2ème Edition. Masson et Cie, Paris, France: 144 p.

Le Floc'h E, 1979 -Elément pour une cartographie de l'occupation des terres en Tunisie aride et saharienne. Project TUN78/007. Programme des Nations Unis pour le Développement "PNUD"/Centre National de Recherches Scientifiques "CNRS", Montpellier, France/Institut des Régions Arides "IRA", Médenine, Tunisia : 27 p.

Le Houérou H.N, 1959 -Recherches écologiques et floristiques sur la végétation de la Tunisie méridionale. Institut de Recherche Scientifique d'Alger, Algeria. Mémoire n° 6. Volume 1: 281 p. Volume 2: 229 p.

Le Houérou H.N, 1969 -La végétation de la Tunisie steppique avec référence au Maroc, à l'Algérie et la Libye. Annales de l'Institut National de Recherche Agronomique, Tunisia. Volume 42, Fascicule 5 : 624 p.

Le Houérou H.N et Ionesco T, 1973 -Appétibilité des espèces végétales dans la zone aride de Tunisie. Document de Travail. Projet FAO/TUN/71/525 : 68 p.

Le Houérou H.N, 1980 -Les fourrages ligneux en Afrique du Nord. In Les fourrages ligneux en Afrique. International Centre for Animal Husbandry in Africa, Addis Ababa, Ethiopia: pp. 57-84.

Le Houérou H.N, 1995 -Bioclimatologie et biogéographie des steppes arides du Nord de l'Afrique. Options Méditerranéennes, series B. Etudes et Recherches. N° 10. Editions Centre International des Hautes Agronomiques Etudes Méditerranéennes: 396 p.

Le Houérou H.N et Pontanier R, 1987 -Les plantations sylvopastorales dans la zone aride de Tunisie. Note technique de MAB. N° 18 : 81 p.

Léone G, 1924 -Consolidamento ed imboschimento delle zone dunose della tripolitana.

L'Agricultura Coloniala, Firense, Il alia. Vol. XVIII (9): pp. 299-308.

Le Tacon F, 1973 -Sol nutrition et production ligneuse. Annales des Sciences Forestières. Volume 30 (3): pp. 259-285.

Long G, 1974 -Diagnostic et aménagement du territoire. Tome I: Principes généraux et méthodes. Masson et Cie Editions, Paris, France: 256 p.

Long G, 1975 -Diagnostic et aménagement du territoire. Tome II: Application au diagnostic phytoécologique. Masson et Cie Editions, Paris, France: 217 p.

Mansouri Lahraoui M, 2011 -Production of Rhizobium inoculum associated with *Acacia saligna* for the revegetation of the Terga quarry (Ain Témouchent). Master. Université Oran 1-Ahmed Ben Bella, Algeria: 80 p.

Maslin B.R and Pedley L, 1982 -The distribution of *Acacia* (*Leguminosae: Mimosaceae*) in Australia. Part. 1. Species distribution maps. Research Notes, Western Australian Herbarium. Vol. 6 : 1-1.

Maslin B.R, 2001a -Flora of Australian. Vol. 11A: *Mimosaceae*, *Acacia*. Part 1. Publication Australian Biological Resources Study, Canberra : 703 p.

Maslin B.R, 2001b -Flora of Australian. Vol. 11B: *Mimosaceae*, *Acacia*. Part 2. Publication Australian Biological Resources Study, Canberra : 562 p.

Maslin B.R, 2002 -**The** role and relevance of taxonomy in the conservation and utilization of Australian Acacias. Conservation Science West Australia 4 (3) : 1-9.

Manjauze A et le Houérou H.N, 1965 -Le rôle des *Opuntia* dans l'économie agricole nord-africaine. Bulletin de l'Ecole National Supérieure d'Agriculture de Tunis. N° 8-9 : pp. 85-164.

Monod T, 1974 -Notes sur quelques *Acacias* d'Afrique et du Proche-Orient. Bulletin de l'Institut de Recherche d'Afrique Noire. Tome XXXVI. Série A. N° 3 : pp. 642-669.

Musset R, 1934 -Le régime pluviométrique de la France de l'Ouest. Bulletin de l'Association des Géographes Français. No. 79: pp. 68-71.

N.A.S, 1980 -Firewood corps: Shrub and species for energy production. National Academy of Sciences: 234 p.

Neffeti M, **Zaâfouri M.S** et Akrimi N, 1990 -Reprise et comportement de quelques espèces arbustives et petites pérennes en milieu aride (Station d'El Fjé, Médenine). Revue des Régions Arides 1/90. Editions Institut des Régions Arides Médenine, Tunisia: pp. 85-112.

Neffeti M Akrimi (N), Floret Ch. et Le Floc'h E, 1991 -Stratégie germinative de quelques espèces pastorales de la zone aride tunisienne. Conséquences pour le semis des parcours. Proceedings IVème Congrès International des Terres de Parcours, Montpellier, France, 22-26/4/1991 : 15 p.

ODS (Office de Développement du Sud, Médenine, Tunisia), 2021 -The Governorates of Gafsa, Gabès, Médenine and Tataouine in figures. ODS website.

Ouled Belgacem A et **Zaâfouri M.S**, 1997 -Impact de la privatisation des terres de parcours collectifs sur la végétation pastorales en zone désertique tunisienne. Options Méditerranéennes, Série A. N° 32. CIHEAM/IAM Montpellier, France. Scientific Editors BOURBOUZE, MSIKA, NASR and ZAAFOURI: pp. 213-2017.

Pardé J, 1961 -Dendrometry. Imprimerie Louis-Jean. Cape: 350 p.

Passavalli F, 1935 -Due importanti *Acacia* da rimboschimonto per le sabbie libiche. L'Agricultura Coloniala, Firense, Il alia. Vol. XXIX (6-7): pp. 299-310.

Peters J, 1974-Inventaire. Publications Ecole Nationale Forestière d'Ingénieurs Salé,

Morocco: 164 p.

Picouet M, Sghaier M et **Zaâfouri M.S**, 1998 -Relation population-environnement en Tunisie désertique. Revue Espace Populations Sociétés. 1998-1 : pp. 53-65.

Poissonet J, Toure I, Gillet H and Cabaret M, 1985 -Methodology for the study of Sahelian pastures. Forum Aménagement Pastoral Intégré au Sahel, Dakar, Senegal: 27 p.

Pottier-Alapetite G, 1979 -Flore of Tunisia. Angiosperms, Dicotyledons. Tome I : Apétales-Dialypétales. Imprimerie Officielle de la République Tunisienne : 652 p.

Pottier-Alapetite G, 1981 -Flore of Tunisia. Angiosperms, Dicotyledons. Tome II : Gamopétales. Imprimerie Officielle de la République Tunisienne : pp. 655-1190.

Rol R, 1954 -Le forestier devant la phytosociologie. Document de l'Ecole Nationale de Génie Rural Eau et Forêt : 25 p.

Saghiri A, 1991 -Sédimentation et morphogenèse au Néogène-Quaternaire dans le Sud-Ouest tunisien. Doctoral thesis. University of Strasbourg. France : 238 p.

Sarson M et Salmon P, 1977 -Composition chimique et valeur alimentaire de certaines plantes forestières spontanées ou introduites au Maroc. Technical Note: 16 p.

Schönenberger A, 1971 -Premiers renseignement des arboretums forestiers en Tunisie. Technical report No. 5. Institut de Reboisement, Tunis/FAO : 178 p.

Schönenberger A, 1986a -Premier rapport phytoécologique sur la forêt du Parc National de Bou- Hedma, et de sa région et l'autoécologie des principales espèces pastorales du Sud tunisien. Tuniso-German Technical Cooperation. Project GTZ/Direction des Forêts. Ministry of Agriculture, Tunisia: 44 p.

Schönenberger A, 1986b -Projet d'aménagement et d'exploitation des ressources naturelles du Parc National du Bou-Hedma et de sa région. Tuniso-German Technical Cooperation. Project GTZ/Direction des Forêts. Minis. Agri, Tunisia. N° 82. 2045. 1-01. 100 : 44 p.

Schönenberger A, 1987 -Deuxième rapport phytoécologique sur les Parcs Nationaux du Sud. Improvement of natural rangelands in Saharan areas and Algerian-Moroccan species likely to be introduced into Tunisia. Tunisian-German Technical Cooperation. GTZ/Direction des Forêts project. N° 82. 2045. 1-01 : 100. 26 p.

Sethom (H), 1979 -Les tentatives de remodelage de l'espace tunisien depuis l'indépendance. Mediterranean, n° 1 et 2 : pp. 119-125.

Tansley A.G, 1935 -The use and abuse of vegetational concepts and terms. Ecology. 16 : pp. 284307.

Tbib A, Chaieb M et **Zaâfouri M.S**, 2000 -**Manifestation** des perturbations écologiques face à l'anthropisation des milieux arides tunisiens : Cas de Menzel Habib. Proceedings of the International Congress MEDENPOP 2000: Population Rurale et Environnement en Contexte Bioclimatique Méditerranéen. Institut des Régions Arides, Médenine/Secrétariat d'Etat à la Recherche Scientifique et Technologique, Tunisia/Institut de Recherche pour le Développement, France. Jerba 25-28 October 2000, 10 p.

IUCN (International Union for Conservation of Nature), 1997 -Red List of Thraetened Plants. K.S. Walter and H.J. Gillett Edition : 682 p.

IUCN (International Union for Conservation of Nature), 1998 -The world list of threatened trees world conservation. S. Oldfield, C.A. Lusty and A. MC Kinver Edition: 650 p.

Vassal J, 1972 -Apport des recherches ontogénétiques et sémiologiques à l'étude morphologique, taxonomique et phylogénique du genre *Acacia*. Bulletin de la Société d'Histoire Naturelle Toulouse.108 (1-2) : pp. 125-247.

Vassal J, 1984-Les *Acacia* (*Mimosa*) en régions méditerranéennes françaises. Compte Rendu de la Société de Biogéographie, France. Volume 16 (1): pp. 19-35.

Visser M, *Nasr N* and **Zaâfouri M.S**, 1997 -Quelle recherche *(agro-) pastorale dans le cadre des mutations agraires?* Options Méditerranéennes, Série A. N° 32. CIHEAM/IAM, Montpellier, France. Scientific Editors BOURBOUZE, MSIKA, NASR and ZAAFOURI : pp. 227-252.

Zaâfouri M.S, 1989 -Floristic and pedological characterisation of a toposequence in the Tunisian arid zone with a view to the installation of fodder shrub species. Preliminary results of trials. D.E.A d'Ecologie des Ecosystèmes Méditerranéens Continentaux. Saint-Jérôme Faculty of Science and Technology, Marseille. Université de Droit, d'Economie et des Sciences d'Aix-Marseille, France: 61 p.

Zaâfouri M.S, Akrimi N, Pontanier R et Zemzemi J, 1991 -Recherches des conditions écologiques optimales de réussite d'arbustes fourragers en zone aride tunisienne : Cas *d'Acacia cyanophylla* Lindl. Proceedings IVème Congrès International des Terres de Parcours. Volume 1. pp. 244-247.

Zaâfouri M.S, 1993 -Contraintes du milieu et réponses de quelques espèces arbustives exotiques introduites en Tunisie présaharienne. Doctoral thesis in Sciences: Biologie des Population et Ecosystèmes. Saint-Jérôme Faculty of Science and Technology, Marseille. Université de Droit, d'Economie et des Sciences d'Aix-Marseille III, France: 202 p.

Zaâfouri M.S, Akrimi N, Le Floc'h E et Zouaghi M, 1994a -Développement d'une approche méthodologique d'évaluation des performances des arbustes fourragers. Options Méditerranéennes. Volume 62. Editions Centre International des Hautes Etudes Agronomiques Etudes Méditerranéennes : pp. 481-484.

Zaâfouri M.S, Akrimi N, Floret Ch., Le Floc'h E et Pontanier R, 1994b -Les plantations sylvopastorales en Tunisie présaharienne. Sécheresse N° 5. Volume 4: pp. 265-275.

Zaâfouri M.S, Akrimi N, Floret Ch, Le Floc'h E et Pontanier R, 1995 -Les arbustes fourragers exotiques. Leur intérêt pour la réaffectation des terres dégradées des régions arides et désertiques tunisiennes. In L'Homme peut-il refaire ce qu'il a défait? John Libbey Editions: pp. 211-229.

Zaâfouri M.S, Nasr N et Jeder H, **1997** -L'attribution des terres collectives et plantations sylvo- pastorales en Tunisie aride et désertiques. Options Méditerranéennes. Série A. N° 32. CIHEAM/IAM Montpellier, France. Scientific Editors BOURBOUZE, MSIKA, NASR and ZAAFOURI: pp. 219-226.

Zaâfouri M.S, 1998 -Inventaire du patrimoine végétal naturel et caractérisation des espèces rares du Sud-Est de la Tunisie. Rapport de Consultation. Ministère de l'Environnement et de l'Aménagement du Territoire. Tunisia/Institut National de Recherche en Génie Rural, Eaux et Forêts. Tunisia: 50 p.

Zaâfouri M.S and Chaieb M, 1999 - Trees and shrubs of southern Tunisia threatened with extinction. Acta Botanica Gallica. Volume 146 (4) : pp. 361-371.

Zaâfouri M.S, Chokri F and Chaieb M, 2016 -Les systèmes écologiques de la région de Gammouda : Etat de connaissance et dynamique au cours de ¾ de siècle (1940-2015). Bulletin des Sciences et Techniques (ISSN 2534-8256). Editions Faculté des Sciences et Techniques de Sidi Bouzid. N° Spécial 2/2016 : pp. 1-78.

Zaâfouri M.S, 2020 -La forêt-Steppe de Gommier de la Tunisie présaharienne à base d'*Acacia tortilis*. Editions Universitaires Européennes (ISBN: 978-620-2-53936-4): 233 p.

Zaâfouri M.S, 2023 -La forêt-steppe d'*Acacia tortilis* (*A. raddiana*) de la Tunisie : Caractérisation, Altération et Dynamique. Edit. Universitaires Européennes (ISBN: 978-620-3-45787-2): 300 p.

CV

Mohamed Sghaïer ZAÂFOURI

-Born on 05 January 1953 in the village of Zaâfria, Sidi-Bouzid, Tunisia
-Residence: Cité Al Wroud, Avenue Yasser Arafat, Sidi-Bouzid.
E-mail address: mszaafouri@gmail.com
-Telephone: (+216) 99217474

Engineer-Doctor-Professor in Higher Agricultural Education

Former Dean and Director General of the Higher Education and Scientific Research Establishments of Sidi- Bouzid, Tunisia (1999-2017)

Graduate of :

-University of Carthage-Tunis, Tunisia :
Habilitation in Agronomic Sciences: Rural, Water and Forest Engineering (2001).
-Université de Droit, d'Economie et des Sciences Aix-Marseille III, France :
PhD in Biological Sciences: Biology of Populations and Ecosystems (1993)
Diplôme d'Etudes Approfondies in Ecology: Continental Mediterranean Ecosystems (1989)
-Ecole Nationale Forestière d'Ingénieurs de Salé, Morocco:
Engineering degree: Water and Forestry (1985)
-Institut Sylvo-Pastoral de Tabarka, Tunisia:
Diplôme de Technicien Supérieur: Forestry (1977)

Teacher-researcher (1985-2018) at the Higher Education and Scientific Research Establishments, Tunisia:

-Sidi-Bouzid Faculty of Science and Technology
-Béja Higher Institute of Technological Studies
-Sidi-Bouzid Higher Institute of Technological Studies
-Sfax Faculty of Science
-Gafsa Faculty of Science
-Kébili Higher Institute of Technological Studies
-Centre-Ouest Regional Agricultural Research and Development Centre (currently the Agricultural Research Centre), Sidi-Bouzid
-Institute of Arid Regions Médenine

Lecturer at the Faculty of Science and Technology of Sidi Bouzid. University of Kairouan, Tunisia (2018-2020)

Director of Research Units :

-Valorisation et Optimisation de l'Exploitation des Ressources *(UR/16 MES 04)*. Faculty of Science and
Techniques of Sidi-Bouzid. University of Kairouan, Tunisia (2015-2018)
-Dynamique des Écosystèmes terrestres en zone aride *(04/UR/09-02)*. Faculty of Sciences of Sfax.

University of Sfax, Tunisia (2009-2012)

Member of a number of research laboratories (1995-2015)

Expert for the General Directorate of Scientific Research. Ministry of Higher Education and Scientific Research, Tunisia.

Member of the Sectoral Commission and the Commission for the Recruitment of Lecturers (2010-2018): Agronomic and Agroindustrial Sciences/Biological Sciences and Ecology.

Founder-Director and Chairman of the Scientific Committee of the Journal: *Bulletin des Sciences et Techniques* (*ISSN 2532-8256*). Edi. Faculté des Sciences et Techniques de Sidi-Bouzid. University of

Kairouan, Tunisia (2015-2020)

Scientific co-editor and member of the Scientific Committees of a number of international and national journals, and international and national conferences (1994-2000)

Coordinator and member of several international and national research projects (1988-2000)

Publications: 2 books + 7 contributions to books + 34 scientific articles in journals, symposia and conferences.

International and national seminars + 5 dissertations (2 theses + 1 post-graduate diploma + 2 final year dissertations)

Carrying out a dozen studies in the field of the environment and sustainable development

Head of General Services at the Institut des Régions Arides, Médenine (1994-1998)

Mayor of the town of Sidi-Bouzid, Tunisia (2005-2010)

Director of the Sidi-Bouzid Cultural Festival, Tunisia (2008-2010)

Member of certain Nature Protection Associations

MIX
Papier aus verantwortungsvollen Quellen
Paper from responsible sources
FSC® C105338

Printed by Books on Demand GmbH, Norderstedt / Germany